Autodesk Inventor Exercises

This practical resource provides a series of Inventor® exercises covering several topics, including:

- sketches
- part models
- assemblies
- drawing layouts
- presentations
- sheet metal design
- welding

for users with some familiarity with Autodesk® Inventor®, or other similar feature-based modelling software such as Solid Works®, CATIA®, Pro/ENGINEER and Creo Parametric, and who want to become proficient. Exercises are set out in a structured way and are suitable for releases of Inventor from versions 7 to 13.

Bob McFarlane has been a Performance Engineer with Rolls-Royce, Curriculum Manager for CAD and New Media at Motherwell College, Scotland, and an Autodesk Educational Developer. He has written over 25 books for AutoCAD users.

Autodesk Inventor Exercises

For Autodesk® Inventor® and Other
Feature-Based Modelling Software

Bob McFarlane

Routledge
Taylor & Francis Group

LONDON AND NEW YORK

First published 2017
by Routledge
2 Park Square, Milton Park, Abingdon, Oxon OX14 4RN

and by Routledge
711 Third Avenue, New York, NY 10017

Routledge is an imprint of the Taylor & Francis Group, an informa business

© 2017 Bob McFarlane

British Library Cataloguing-in-Publication Data
A catalogue record for this book is available from the British Library

Library of Congress Cataloging in Publication Data
Names: McFarlane, Robert, author.
Title: Inventor exercises : for Autodesk Inventor and other feature based modelling software / Bob McFarlane.
Description: Milton Park, Abingdon, Oxon ; New York, NY : Routledge, 2017. | Includes bibliographical references and index.
Identifiers: LCCN 2016046309| ISBN 9781138849181 (pbk. : alk. paper) | ISBN 9781315725802 (ebook)
Subjects: LCSH: Engineering graphics—Data processing—Problems, exercises, etc. | Engineering models—Data processing—Problems, exercises, etc. | Autodesk Inventor (Electronic resource)
Classification: LCC T386.A974 M34 2017 | DDC 620/.0042028566—dc23
LC record available at https://lccn.loc.gov/2016046309

ISBN: 978-1-138-84918-1 (pbk)
ISBN: 978-1-315-72580-2 (ebk)

Typeset in Bembo
by Florence Production Ltd, Stoodleigh, Devon, UK

Visit the companion website: www.routledge.com/cw/mcfarlane

Contents

vi *Contents*

About the author

Bob McFarlane was born in Renfrew, a town to the south-west of Glasgow. He was educated at Renfrew High Junior Secondary and then at Paisley Camphill Senior Secondary, where he obtained the Scottish Sixth Year Certificate in Mathematics (Analysis, Geometry and Dynamics) and Higher grade passes in Mathematics (A), Physics (A), Chemistry (A) and English (C). He also obtained O grade/Lower passes in History, Geography, French, Science, Arithmetic and Applied Mechanics. On leaving school, Bob obtained a Mechanical Fitting apprenticeship with the South of Scotland Electricity Board (SSEB) and was based at Braehead Power Station in Renfrew.

Bob won an industrial scholarship to the University of Glasgow, where he obtained a 1st class Honours degree in Mechanical Engineering, his final year subjects being Thermodynamics, Structures, Engineering Design, Nuclear Power and Mathematics. He also obtained a 1st class ARCST from the Royal College of Science and Technology (now the University of Strathclyde). Bob's final-year thesis was in the field of Thermodynamics and entitled 'Investigating the Reynold's number effect of inclined condenser tubes'.

On leaving university, Bob obtained employment with Rolls-Royce as a Performance Engineer, first at Hamilton and then at the East Kilbride plant. His specialist areas were compressor surge problems and turbine blade design, and he worked on several turbo-jet and turbo-prop engines.

In the early 1970s, Bob made a career move into education, first as a Secondary School Technical Teacher at Lochend and Garthamlock Secondary Schools in the east end of Glasgow, and then as a lecturer in Further Education, where he taught traditional Engineering/Mathematics topics to apprentice engineers from several well-known companies. It was in FE that Bob became interested in CAD, and used his first CAD package in 1983–1984 (AutoCAD Release 2.5) with no colour and no solid modelling (i.e. basic 2D CAD). This was in the early days of CAD in education, and Bob had the foresight to realise that CAD had a future in both industry and education, and he thus developed and pioneered the HNC/HND courses in Computer Aided Draughting and Design (CADD), the first course of its type in the world. These courses are still running today in many colleges throughout Britain and offer students considerable employment prospects, as well as a route into many university courses. While developing these courses, Bob returned to university as a mature student and obtained his Masters (MSc) in Computer Integrated Manufacture and submitted his PhD thesis on 'Modelling techniques using three-dimensional data obtained from co-ordinate measuring equipment and customised programming'.

Bob has published over 25 books for the AutoCAD Draughting Package, covering topics such as 2D draughting, 3D draughting, solid modelling, customisation, AutoLISP programming, etc. These books covered releases from R10 to AutoCAD 2007.

Bob retired from full-time education in 2008, but is still involved with CAD, being an online tutor for the ICS correspondence AutoCAD courses, as well as a Work-Based Assessor for apprentice draughtspersons with a large well-known multinational company.

Professionally, Bob is a Chartered Engineer (CEng), a Registered Engineering Designer (REngDes), and a Fellow of the Institution of Engineering Designers (FIED), as well as having membership of several other institutions. Bob is also the Scottish Regional Co-Ordinator for the IED.

In his spare time, Bob is a keen distance swimmer and an avid stamp collector for New Zealand and Great Britain.

Bob lives very happily in Bellshill with his wife Helen and has two children, Linda and Stephen, and one granddaughter, Ciara Erin Docherty.

Acknowledgements

The models in the exercises are not all my own thoughts and ideas.

I used several engineering and draughting books and selected those exercises that would be interesting and challenging to the user. The following authors must be acknowledged:

Alf Yarwood	*Introduction to AutoCAD 2013*
Dennis Maguire	*Engineering Drawing from First Principles Using AutoCAD*
Pickup and Parker	*Engineering Drawing with Worked Examples*

Without the vision of the above authors, it would not have been possible for me to compile this series of exercises, so my sincere thanks to all of them.

Introduction

Having used Inventor for several years, I always wanted to attempt new models, and searched various textbooks etc. for ideas, and am sure most readers will have done the same. I then decided that a book of exercises would be very useful to all feature-based modelling users who wanted additional models to attempt.

The book is *not* intended as a teaching manual, as I assume that readers have FBM experience and know how to use Inventor® (or other software) for their own requirements. The book is, as the title says, a series of Inventor exercises covering several Inventor 'topics', including:

- sketches;
- part models;
- assemblies;
- drawing layouts;
- presentations;
- sheet metal design; and
- welding.

Hopefully, the reader will work through the exercises (which cover a wide range of engineering and other disciplines) to become even more proficient at using Inventor.

The exercises are suitable for:

(a) *all* Inventor releases (I completed them using Inventor 7 and checked several with other releases); and
(b) other feature-based modelling software (e.g. Solid Works, CATIA, Pro-Engineer, etc.).

I have to acknowledge the help from former colleagues for allowing me to use their ideas in these exercises.

Bob McFarlane, October 2016

File types

The four file types that are used with the exercises are:

- .ipt – part models;
- .iam – assemblies;
- .idw – drawing layouts; and
- .ipn – presentations.

A 2D sketches

1 In this first series of exercises, you have to create 2D sketches.

2 These sketches should then be constrained, dimensioned and 'converted' into a part model using one of the 3D part feature tools (e.g. extrude, revolve, sweep, etc.).

3 In each exercise, the procedure for creating the model is:
   (a) Start each exercise with a new metric standard (mm) .ipt file.
   (b) Complete the sketch using the information displayed.
   (c) Fully constrain the sketch.
   (d) Add the dimensions as given (or to your requirement).
   (e) Extrude or revolve the dimensioned and constrained sketch.
   (f) View at a suitable 3D viewpoint.
   (g) Save your completed part model to a suitably named folder.
   (h) Use discretion when creating the model, as appropriate.
   (i) *Note*:
       • Some of the sketches are relatively simple, but I have also included others to make you 'think a bit'.
       • If you are relatively proficient at creating part models, then you may want to miss this chapter, although it is still good practice – this is your decision.
       • In the exercises, I will display the 2D sketch with dimensions and the created part model from the sketch.

Terminology

Sketch
1 A sketch is a plane on which 2D objects are drawn (i.e. sketched).
2 The plane can be XY, YZ, XZ or a user-positioned work plane.

Constraints
1 Geometric constraints apply behaviour to a specific object or create a relationship between two objects.
2 For example, a line may be constrained to be horizontal or two lines may be constrained to be equal in length.

Dimensions
1 The dimensions 'added' to a sketch control the size of the sketch and will be displayed in the drawing view when generated.
2 Inventor dimensions are *parametric* (i.e. if a dimension value is altered, the shape of the object (model) will also be altered).

Base feature
The first sketch of a part that is to be used to create a 3D feature is referred to as the base feature.

3D feature
When a sketch has been extruded or revolved etc., the result is termed a 3D or part feature.

Exercise A1: shim

The sketched profile has to be extruded for a distance of 15mm.

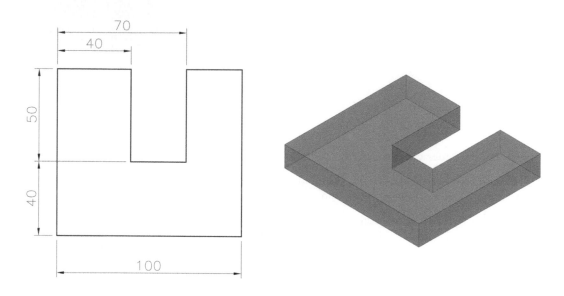

Exercise A2: spacer

The dimensioned sketch has to be extruded for a distance of 20mm.

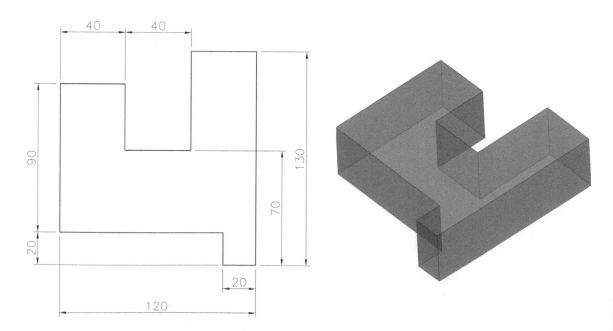

Exercise A3: lock guide

Extrude the sketch for a distance of 18mm.

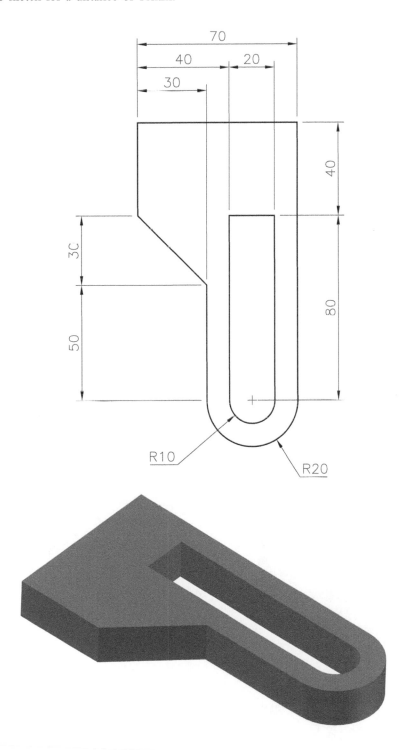

Exercise A4: rocker arm

Extrude the sketch for a distance of 10mm.

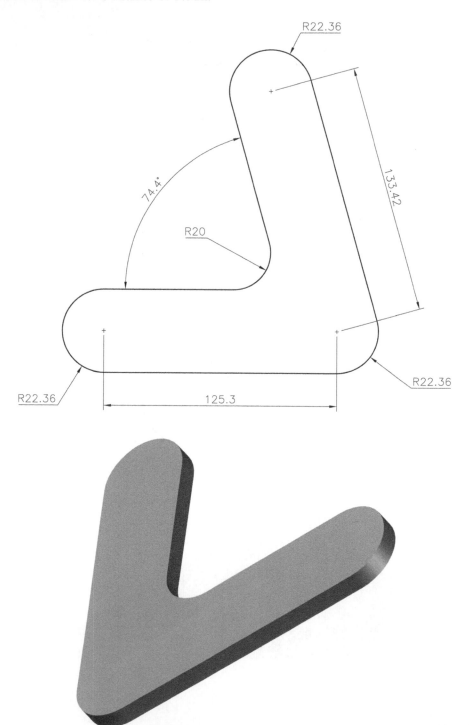

Exercise A5: clip

Extrude the sketch for a distance of 5mm.

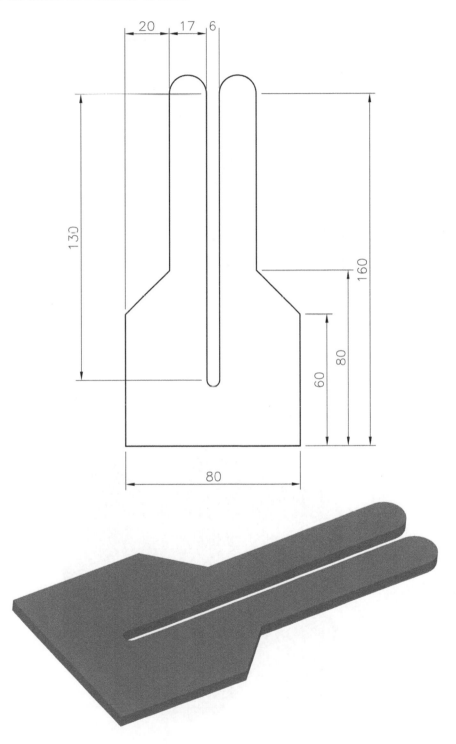

Exercise A6: V block

Extrude the sketch for a distance of 180mm.

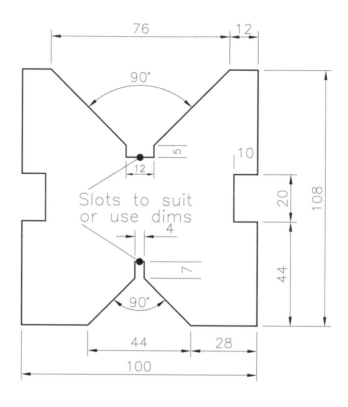

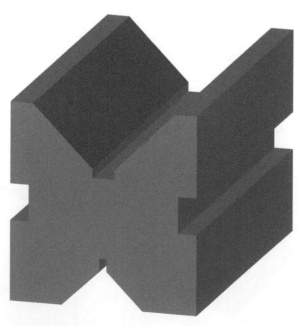

Exercise A7: coupling link

Extrude the sketch for a distance of 11mm.

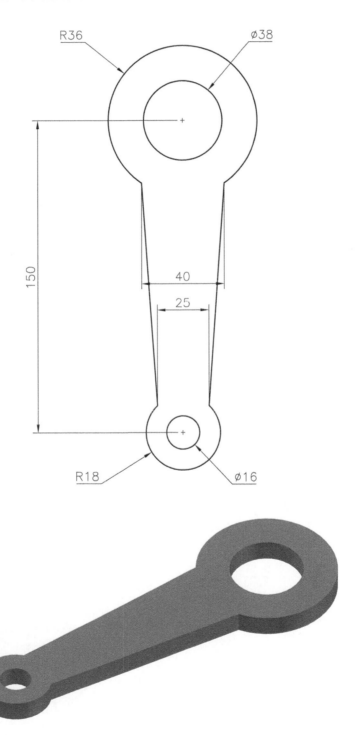

Exercise A8: container

Extrude the sketch for a distance of 120mm.

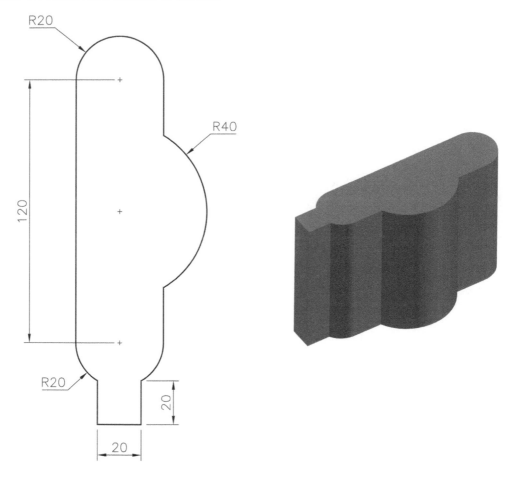

Exercise A9: support

Extrude the sketched feature for a distance of 6mm.

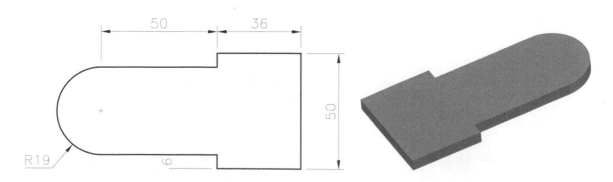

Exercise A10: pulley wheel

Revolve the sketched feature for 270 degrees then rotate to suit.

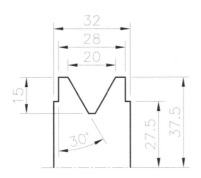

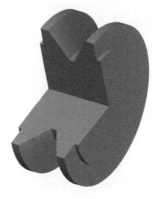

Exercise A11: template

Extrude the sketch 15mm.

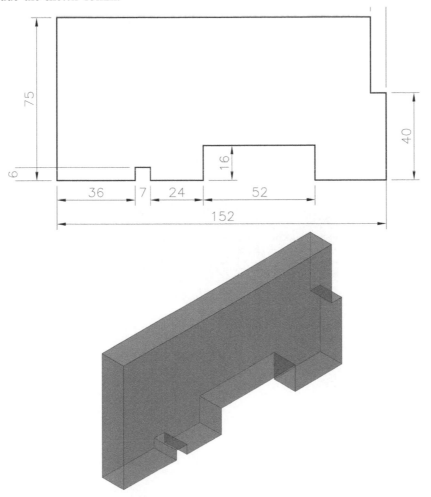

Exercise A12: block

Extrude the block sketch for 50mm.

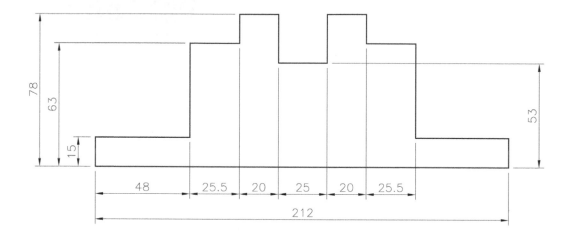

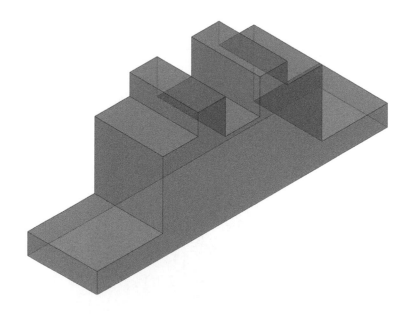

Exercise A13: rocket

Two part models to be created from the one sketch:

1 extrude for 25mm; and

2 revolve for a complete revolution.

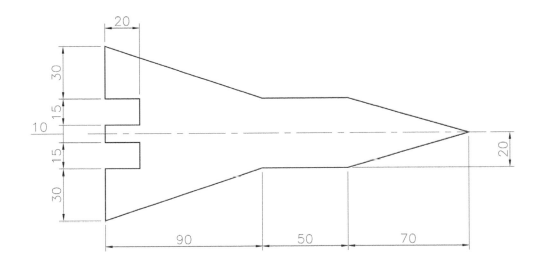

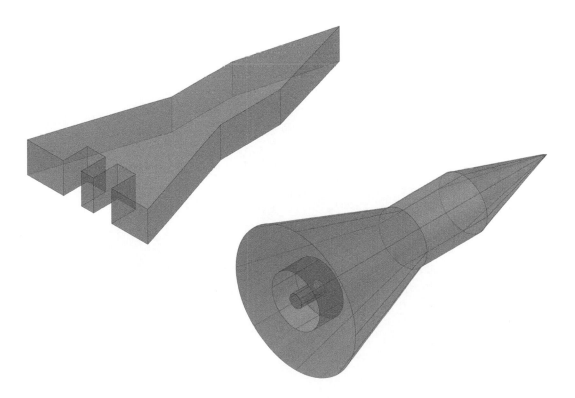

Exercise A14: component

Extrude the dimensioned sketch for 15mm.

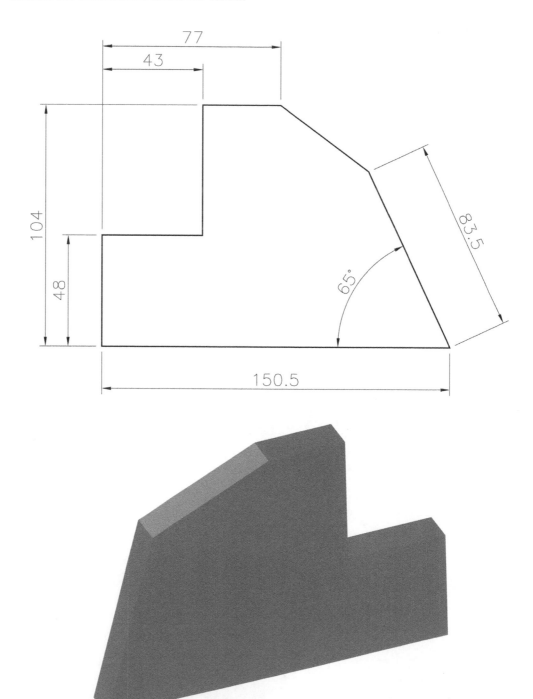

Exercise A15: tree

Two part models to be created from one sketch:

1   extrude for 10mm; and

2   revolve for a complete revolution.

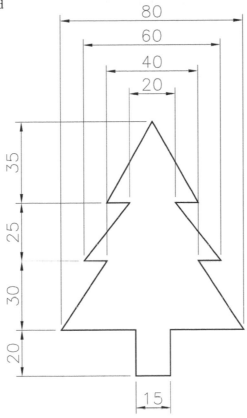

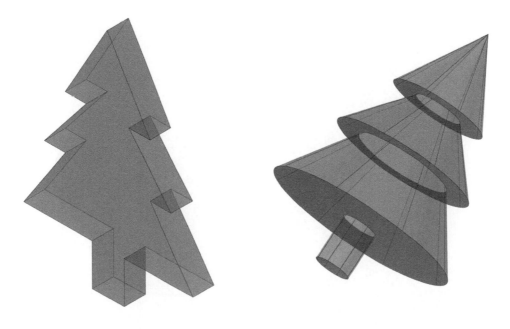

Exercise A16: cam

The dimensioned shape is to be extruded for 20mm.

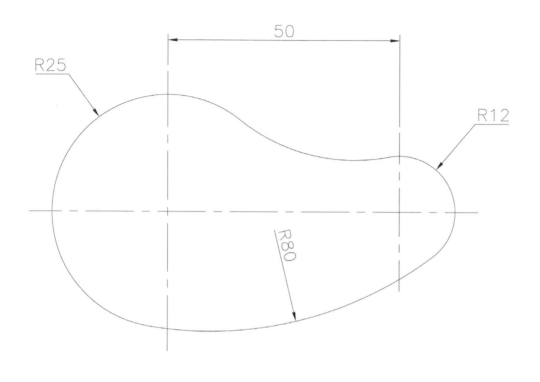

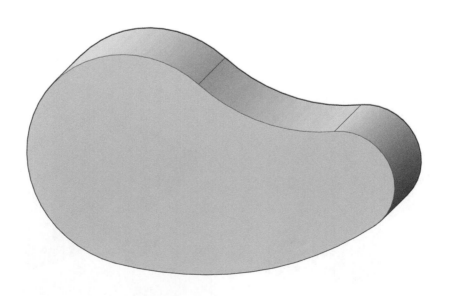

Exercise A17: plane handle

Some interesting work to complete the sketch, which is to be extruded for 30mm.

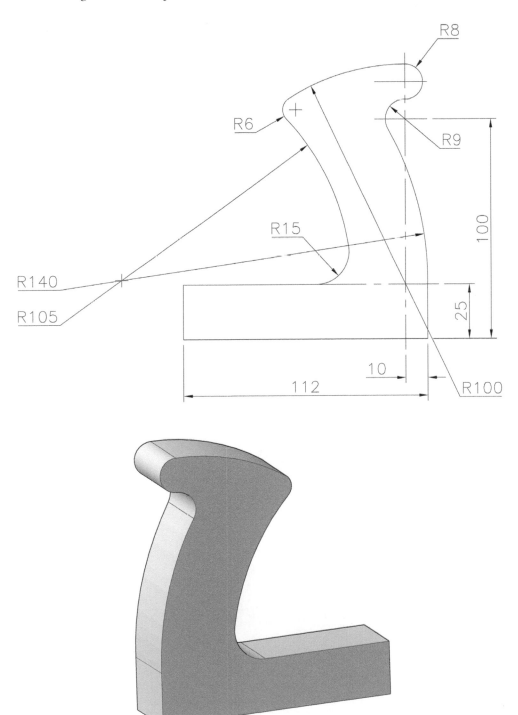

Exercise A18: spanner end

Extrude the sketch for a distance of 12mm.

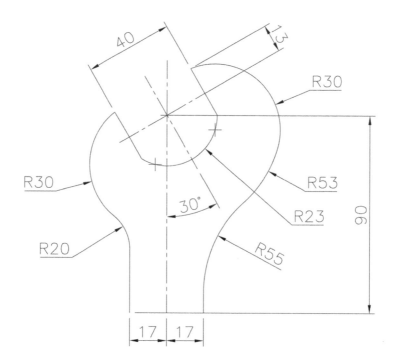

Exercise A19: metalwork dolly

Extrude the sketch (which requires an ellipse) for a distance of 15mm and revolve for a complete revolution.

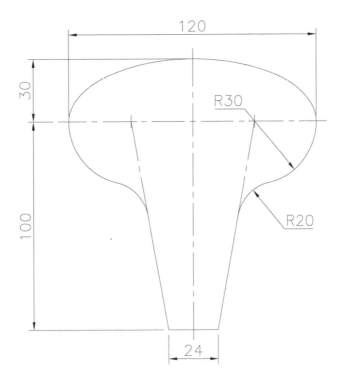

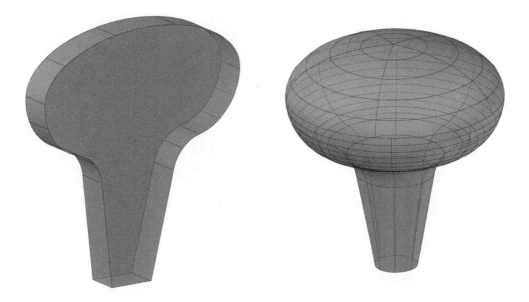

Exercise A20: crane hook

Extrude the dimensioned sketch of the crane hook, selecting a suitable extruded distance (this is really a tangency problem).

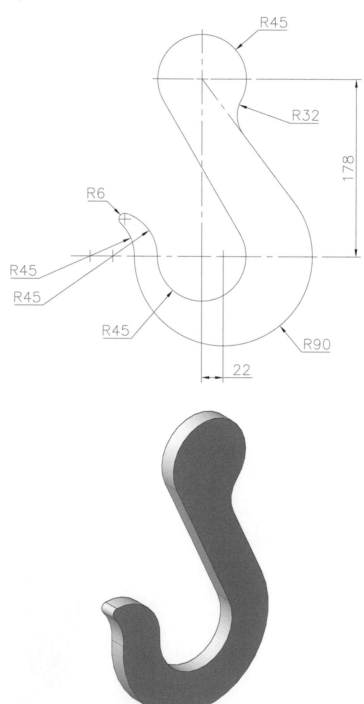

B Placed features 1

1   With the A exercises, the user created dimensioned, constrained sketches and extruded/revolved these sketches into Inventor part models.

2   In this second series of exercises:

(a) original sketches/part models will be modified to include part features; and

(b) new sketches/part models with part features will be created.

3   Inventor® lists the following as part features:
(a) fillet;
(b) chamfer;
(c) hole;
(d) thread;
(e) shell;
(f) decal;
(g) emboss; and
(h) face draft.

4   Part features can be 'added' to the original sketch or the part model itself, and the exercises in this chapter will only consider the fillet, chamfer, hole and thread part features.

5   The other part features will be considered in the next chapter.

6   The process for the exercises is:

(a) *Modifying saved part models*
- Start each exercise with *an already completed part model* file.
- Modify the original sketch or part model as required.
- Ensure that the modified sketch is still fully constrained.
- Add dimensions if necessary.
- Obtain the modified part model and view at a suitable 3D viewpoint.
- Save your completed modified part model to your named folder (with a new name).
- Use discretion as appropriate.

(b) *New part models*
- Start each exercise with the *standard metric (mm) .ipt* file.
- Create the dimensioned, constrained sketch.
- Obtain the part model and view at a suitable 3D viewpoint.
- Save your new part model to your named folder.
- Use discretion as appropriate.

Terminology

Sketched feature

Where the user draws a sketch on a planer face (or work plane) and either adds or subtracts material to or from existing features in a part.

Tools

Use the Extrude, Revolve, Sweep or Loft tools to create sketched features in a part.

Placed feature

1   Features that are predefined and only need to be located.
2   The placed features considered (as stated) are fillet, chamfer, hole and thread.

Fillet

Fillet features consist of fillets and rounds, and:

1   a fillet adds material to interior edges to create a smooth transition from one face to another;
2   rounds remove material from exterior edges; and
3   fillet options are:
    (a) Edge: allows individual selected edges to be filleted.
    (b) Loop: fillets all edges that form a closed loop with the selected edge.
    (c) Feature: will fillet all edges of a selected feature.
    (d) All fillets: will select all concave edges of a part that are not filleted.
    (e) All rounds: will select all convex edges of a part that are not filleted.

Chamfer

Similar to fillets except that the edge is bevelled rather than rounded, and:

1   when a chamfer is created on an interior edge, material is added to the model;
2   when the chamfer is added to an exterior edge, material is removed; and
3   chamfer options include:
    (a) Distance: creates a 45-degree chamfer on a selected edge.
    (b) Distance and angle.
    (c) Two distances.

Holes

1   The name says it all, and Inventor has four types of hole available to the user, these being:
    (a) straight through or blind;
    (b) counter-bored;
    (c) counter-sunk; and
    (d) threaded.

2   There are three 'basic steps' for creating holes:
    (a) Create a new sketch on which the hole will exist.
    (b) Create point hole centres to represent the centre of the hole.
    (c) Use the Hole tool and Hole dialog as appropriate.

Threads

1   Thread features are used to create both internal threads (holes) and external threads (shafts, studs, bolts).

2   The threads are displayed on the parts with a 'graphical representation' (i.e. they *do not physically exist* on the part).

3   With threads, the following parameters are available for user selection:
   (a)   thread location;
   (b)   thread length;
   (c)   offset;
   (d)   direction;
   (e)   type;
   (f)   nominal size;
   (g)   class; and
   (h)   pitch.

Suggestion: It is *recommended* that, where possible, holes, fillets and chamfers are 'added' to the completed part model – think about this suggestion.

A. Modifying existing sketches/part models

Exercise B1: modified spacer (A2)

The original part model has to be modified:

1   to include the part features indicated; and

2   by extruding for a distance of 17 with a 3-degree taper.

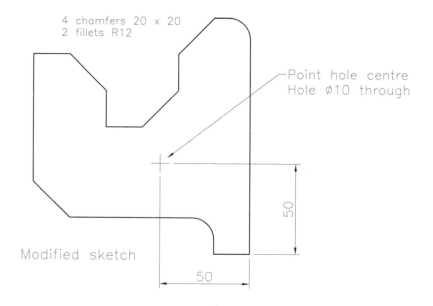

4 chamfers 20 x 20
2 fillets R12

Point hole centre
Hole Ø10 through

Modified sketch

50

50

Exercise B2: modified rocker arm (A4)

1   The rocker arm has been redesigned to be 28mm 'thick' with three holes positioned on the 'curve centres'.

2   The information to create the holes is:
  (a) Holes 1 and 2: full depth ISO M16 thread.
  (b) Hole 3: ∅8 through with a counter-bore ∅24 for a depth of 8.

3   To assist with strengthening, a 30 × 30 chamfer has been added, as shown.

4   Create the rocker arm with the modifications listed.

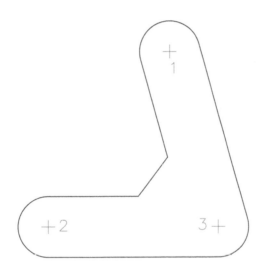

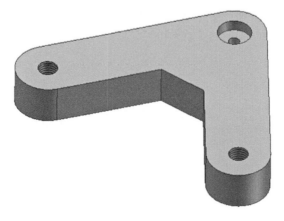

Exercise B3: modified coupling link (A7)

The original component has to be re-extruded for a distance of 18mm and four R15mm fillets have to be added. These additional features have resulted in the component being too heavy, and the following measures have had to be introduced:

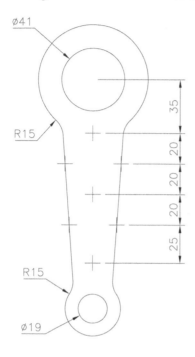

1   The circle diameters have been increased to 41 and 19.
2   Seven Ø10 holes have been included, the point hole centres being as shown.

Create the new part model with the modifications included and save.

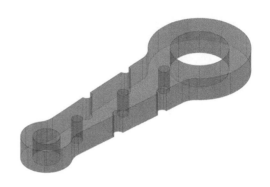

Exercise B4: modified container (A8)

1   Modify the original extrusion to that shown and add the three holes using the information provided.
2   The hole point centres are positioned on the curve centre points.
3   The final model extruded distance is 40mm.

Exercise B5: modified support (A9)

Use the new information given to create a modified support plate.

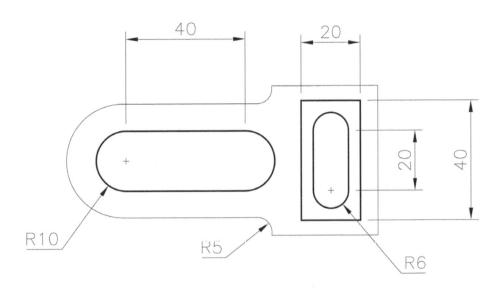

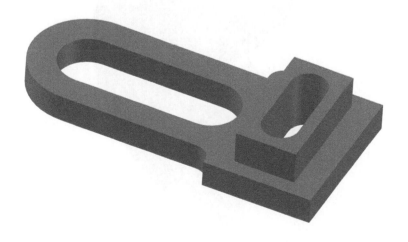

Exercise B6: modified pulley wheel (A10)

1   Modify the existing revolved part model profile to that displayed below.

2   Revolve the 'new' profile for 230 degrees.

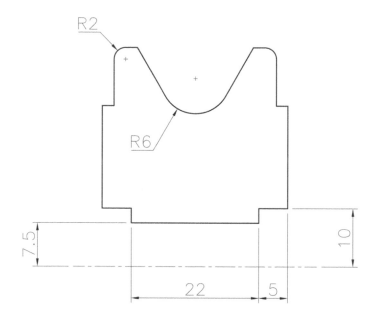

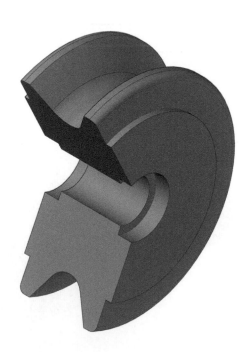

Exercise B7: modified template (A11)

The existing extruded part model has to be altered to include the modifications, as shown below, and then extrude for 27mm.

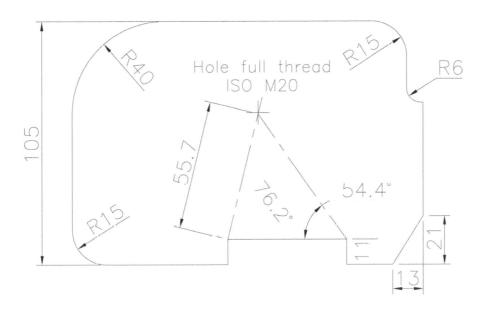

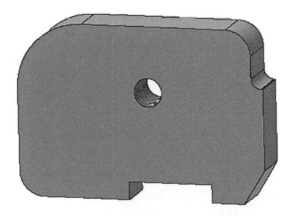

Exercise B8: rocket (A13 extruded)

1   Modify the extruded rocket part model to include the 10 fillet radii as displayed.

2   Extrude the modified rocket for a new distance of 16mm.

3   Five different-sized holes have to be added 'along the rocket centre line', the details being given below.

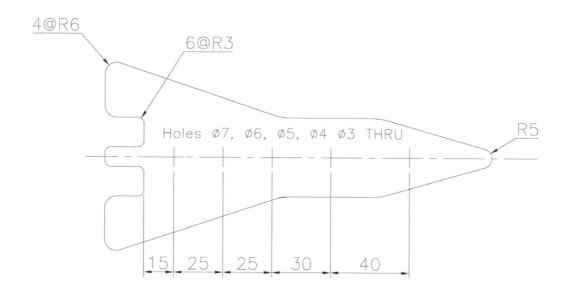

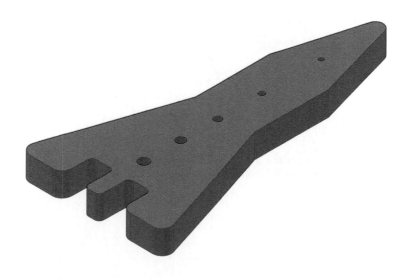

Exercise B9: cam (A16)

1   The original extruded cam requires two holes to be included, the details being given below.

2   The R80 curve has to be replaced with a straight line to assist with the linear motion for the cam follower.

3   Incorporate these changes to create a new cam part model.

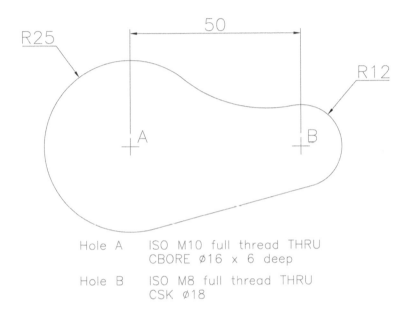

| Hole A | ISO M10 full thread THRU CBORE ø16 × 6 deep |
| Hole B | ISO M8 full thread THRU CSK ø18 |

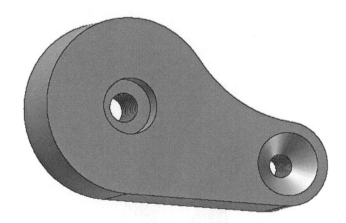

Exercise B10: crane hook (A20)

Create a new part model of the crane hook, which has been modified to include:

1   a 5mm boss added to both sides of the 'top end';

2   an ISO M40 threaded hole included in the 'new boss';

3   the new boss chamfered 3 × 3 both sides;

4   an R4 fillet added to both sides of the 'hook body'; and

5   the 'free end' now with a radius of 10mm.

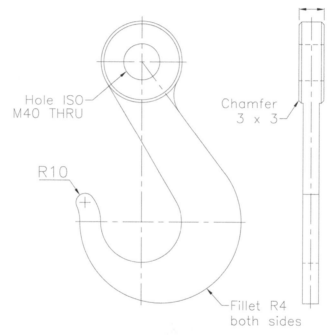

B. New part models

Exercise B11: bottle opener

Using the orthographic drawing information, create a part model of the bottle opener.

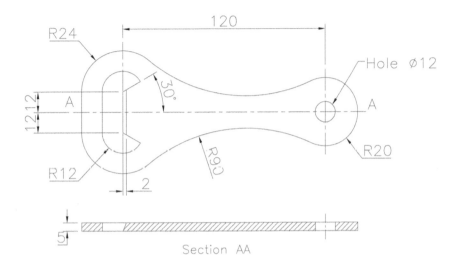

Section AA

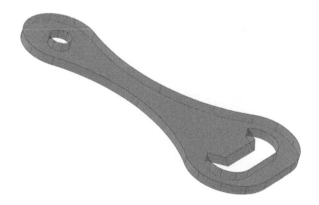

Exercise B12: copper plate

1   Two orthographic views of the copper plate are displayed.

2   Create a part model of the plate.

3   This model will be referred to in a later section.

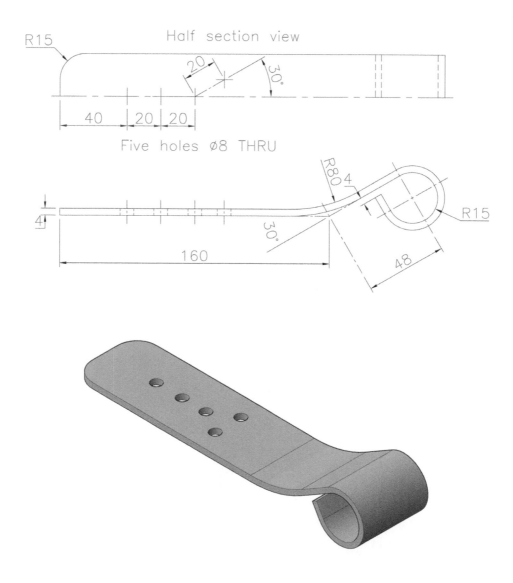

Exercise B13: signal arm

Create a part model of the signal arm that has a thickness of 9mm.

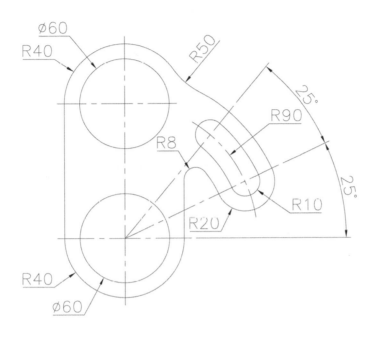

Exercise B14: gasket

The drawing below is for the creation of a blue rubber 4mm thick gasket.

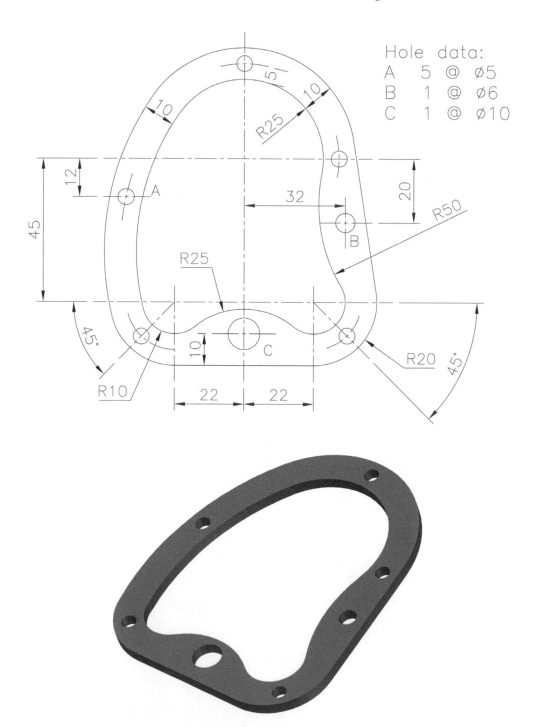

Hole data:
A 5 @ ⌀5
B 1 @ ⌀6
C 1 @ ⌀10

Exercise B15: earthenware vase

1   The outline of the earthenware vase has to be extruded for 18mm and also revolved for a complete revolution.

2   Create the two resulting part models.

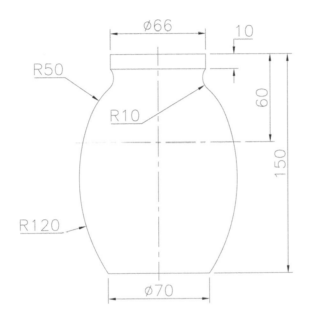

Exercise B16: spectacles

1   This exercise is a bit different from previous exercises.

2   There are three steps given in the creation of the 5mm white plastic spectacles.

3   The information given should be easy for you to understand.

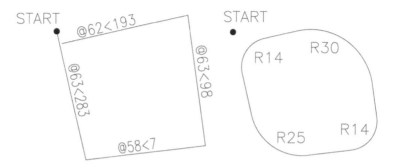

Step 1: the lines   Step 2: the fillets

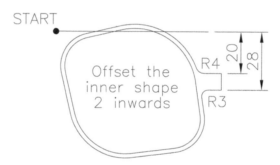

Step 3: one complete 'eye-piece'

Exercise B17: *template*

A 4mm thick template has to be created from red clear material using the reference sizes displayed below.

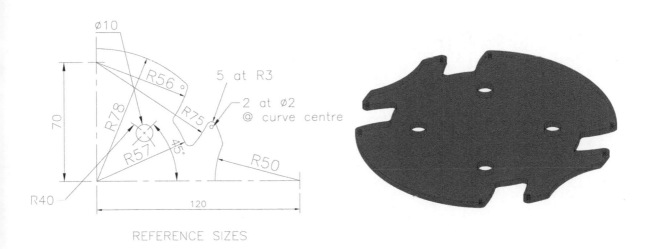

REFERENCE SIZES

Exercise B18: *bicycle spanner*

Create a 7mm extruded part model of the bicycle spanner using the drawing details below.

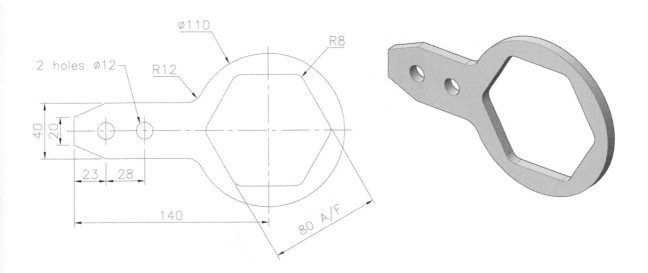

Exercise B19: *fixing bracket*

1   Use the model drawing below to create a part model of the bracket.
2   The ISO M6 holes are positioned 10mm from the front vertical face of the displayed model.

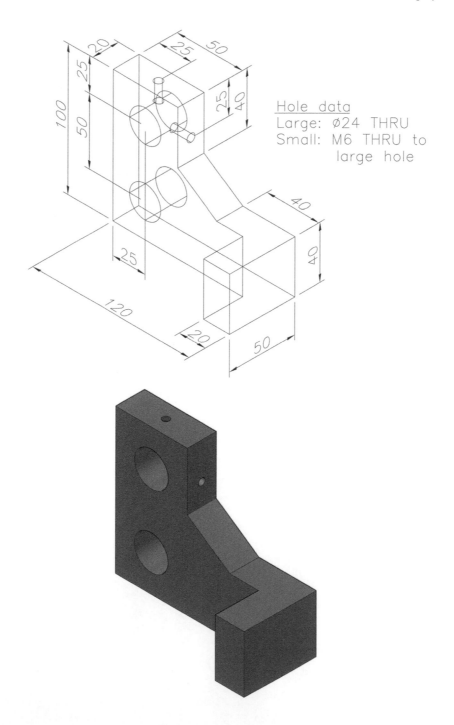

Hole data
Large: Ø24 THRU
Small: M6 THRU to
          large hole

Exercise B20: *curved clip*

1   Use the model of the clip drawing below to create a part model of the bracket.

2   The ISO M6 holes are positioned on the 'clip centre line'.

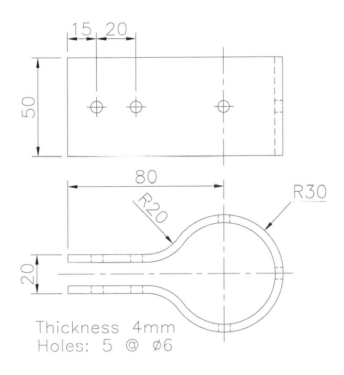

C Placed features 2

1   The placed features investigated in the B exercises were fillets, chamfers, holes and threads.

2   The C exercises will cover the shell, emboss, decal and face draft placed features.

3   The procedure for the various exercises is:
    (a) *For existing models*
         •   Open the named (and saved) part model.
         •   Modify the existing sketch/model using the information given.
         •   Create the new part model and view at a suitable 3D viewpoint.
         •   Save the new part model to your named folder.
         •   Use discretion as appropriate.

    (b) *For new models*
         •   Start each exercise with a new *standard metric (mm) .ipt* file.
         •   Using the drawing and information given, sketch, constrain and dimension.
         •   Create the part model from the fully constrained sketch.
         •   View at a suitable 3D viewpoint.
         •   Save your completed part model to your named folder.
         •   Use discretion as appropriate.

Terminology

Shell
The term refers to removing material from the interior of a part, thus creating a hollow 'container' having a specified wall thickness.

Emboss

1   An embossed feature is created by raising or recessing a profile relative to the model face, the user specifying the depth and direction of the recess.

2   The 'embossed area' can be used as a surface for a decal.

3   Options are:
    (a) emboss from a face;
    (b) engrave from a face; and
    (c) emboss/engrave from a plane.

Decal

1 A decal feature can be created by applying an image to a part face (e.g. a company logo).

2 The images that can be used include bitmaps and Word docs.

Face draft
Used to apply a 'tapered angle' to a selected face using a draft angle.

Exercise C1: rocker arm (B2)

1 Open the modified rocker arm part model, saved as B2.
2 Change the model orientation to that displayed below.
3 Shell by removing the face indicated for a 2mm thickness.
4 Save the shelled model to your named folder.

Exercise C2: container (B4)

1 Open the modified container part model from Exercise B4.
2 Delete, from the model hierarchy, the three created holes.
3 Create 'shelled' part models:
   (a) removing the top face with a thickness of 3mm; and
   (b) producing a 'hollow effect' with a wall thickness of 5mm.
4 Save both models as appropriate.

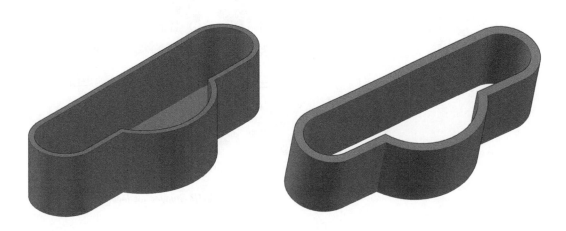

Exercise C3: tree (A15)

1   Open the original part model of the extruded tree.
2   Emboss for 3mm suitable text 'onto' one of the tree faces.
3   'Colour' the model/text to suit.
4   Save when complete.

Exercise C4: creating a company logo

1   We want a company-type logo that has been created and saved as a *BMP* file.
2   This BMP file can then be used with the *DECAL* feature.
3   If you have access to a 'suitable' BMP file, then it is your decision whether to complete the exercise or not.
4   Below is my logo idea, displayed as:
   (a)   an AutoCAD drawing; and
   (b)   possible BMP logo 'sizes' for use with DECAL (probably use Logo 8).

(a)   *AutoCAD logo*

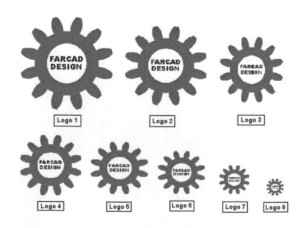

(b)   *BMP 'images' of logo*

Exercise C5: bottle opener (B11)

1 Open the original part model of the bottle opener.
2 Emboss and engrave suitable text to the model face indicated, the emboss and engrave 'distances' being 1mm.
3 'Colour' the model/text to suit (if appropriate).
4 From the 2D sketch panel, insert an image of your logo 'onto' a suitable surface, then use the part features DECAL command, selecting the inserted image.
5 Save when complete.

Exercise C6: plane handle trophy (A17)

1 Open the original part model of the plane handle.
2 Edit the extrusion feature for a distance of 50mm.
3 Add a face draft of 5 degrees to the 'front' vertical surface with a 'backwards' direction.
4 Ensure you select all relevant faces for the draft and note that the face draft has been added; to the outside surfaces if the part model.
5 Emboss suitable text for 1mm onto the front vertical surface.
6 Decal your logo image onto the front vertical surface.
7 Apply a suitable material then save when complete.
8 Note that the part model display below has been rotated so that the face draft effect can be visualised.

Exercise C7: copper plate (B12)

1   Open the original part model of the copper plate.
2   Create suitable text on an appropriate sketch plane that will allow the text to be engraved from the sketch plane 'through' the copper plate.
3   Now insert your logo BMP image (or other suitable image) onto three different surfaces as displayed:
   (a) top horizontal;
   (b) side vertical; and
   (c) end curved.
4   Decal the inserted images onto the surfaces selected.
5   Save the completed model with the Emboss and Decal features.

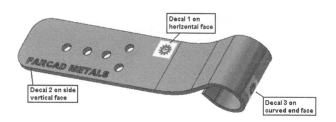

Exercise C8: metalwork dolly (A19)

1   Open the original part model of the extruded metalwork dolly.
2   Edit the extrusion feature and alter the distance to 30mm.
3   Change the orientation from a vertical to horizontal position.
4   Shell the top surface for 4mm.
5   Face draft the outside faces for 5 degrees.
6   Save as required.

Exercise C9: earthenware vase (B15)

1 Open the original part model of the revolved earthenware vase.
2 Shell the vase for 5mm.
3 Use the SPLIT feature tool and remove the 'front half' of the vase.
4 Save.

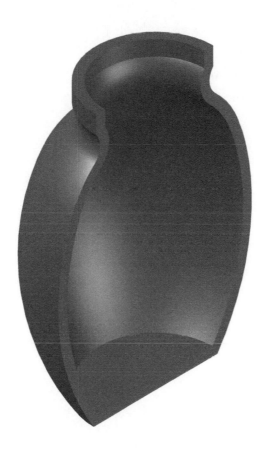

Note: **The exercises that follow are all new, so start each exercise with the standard metric (mm) .ipt file.**

Exercise C10: container

Using the sketch details below, create an extruded part model of the container with the following 'parameters':

1   Extrusion distance: 40mm.

2   Shell thickness: 5mm.

3   Face draft: 2.5mm all inside vertical faces.

4   Fillet: 2mm all inside vertical and horizontal edges.

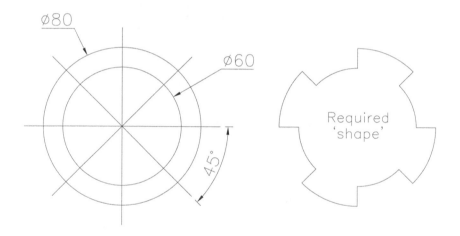

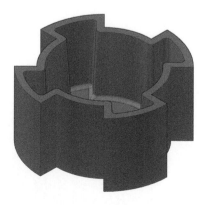

Exercise C11: polygon pyramid

A pyramid has to be created from regular polygons using the following information:

1   There are eight 'steps' in the pyramid.

2   Each step is 20mm in height and has the shape of a regular polygon.

3   The base polygon has 10 sides and is circumscribed in a 100mm circle.

4   The second step in the pyramid has nine sides and is circumscribed in a 90mm circle.

5   The third step has eight sides and is circumscribed in an 80mm circle.

6   And so on, until the top tier has three sides, circumscribed in a 30mm circle.

7   The name of each step in the pyramid has to be embossed onto one of the vertical faces for 1mm and coloured.

Now create the part model of the pyramid.

Exercise C12: embossed metal key ring

1   The local bird watchers club has asked members for two designs, these being:
    (a)   a new logo that will be used on future publications; and
    (b)   a shiny new key ring.

2   Using the details below as a guide, design (and save) a new BMP logo.

3   Create a part model of the key ring 3mm thick with suitable text engraved onto the top face
    and the new logo inserted as text and then 'decaled' onto the top face.

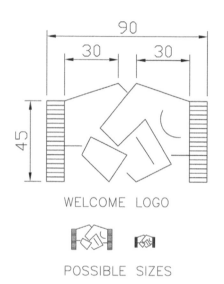

WELCOME LOGO

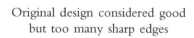

POSSIBLE SIZES

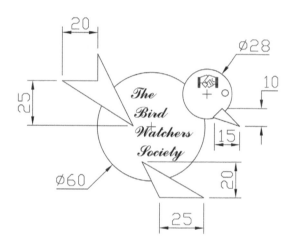

Original design considered good
but too many sharp edges

Modified design accepted

Exercise C13: brass hanger

1   Create a brass hanger from the X-section detail below.

2   Emboss the company name on the side.

3   The hanger is 15mm wide and the embossed text is 0.2mm 'high'.

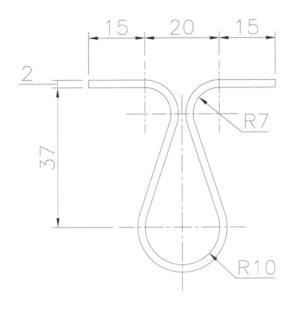

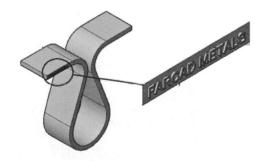

Exercise C14: drain plate cover

1 Create a part model of the drain plate cover using the information from the two First Angle projection views and *save* for future recall in the E exercises.

2 *Suggestion only:*
   (a) Extrude the basic outline shape for 10mm.
   (b) Shell for a thickness of 1.5mm.
   (c) Add a 10-degree face draft to the inside faces.
   (d) Add the internal R3 fillet – six places.
   (e) Add one external R3 fillet.
   (f) Add a ∅5mm hole using the dimensions given.
   (g) Emboss suitable text (height 0.5mm) on a suitable face.

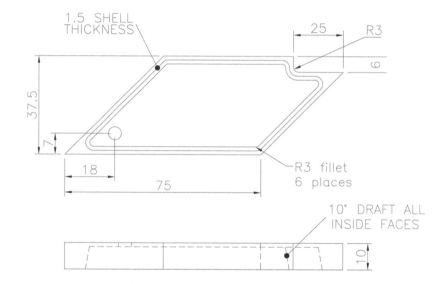

Exercise C15: connector

Suggestion only:

1   Create the main part of the model.

2   Include a 10-degree face draft on the ∅20 curved surface.

3   Add the 15mm 'extension' part and the two thread features.

4   Suitable embossed text is to be added onto a selected face.

5   Save the completed part model for the E exercises.

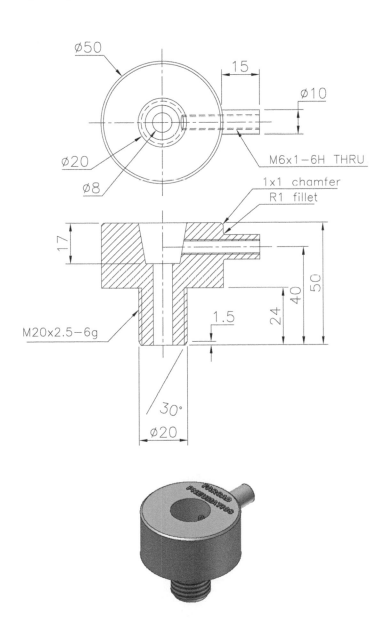

Exercise C16: lettered cube

1   Create a 50mm coloured cube with a 2mm fillet on all edges.
2   The letters of FARCAD have to be embossed for 1mm on the six faces.
3   The size and orientation of the letters is at your discretion.
4   Remember to save for future recall.

Exercise C17: container

Use the drawing information to create a part model of the container.

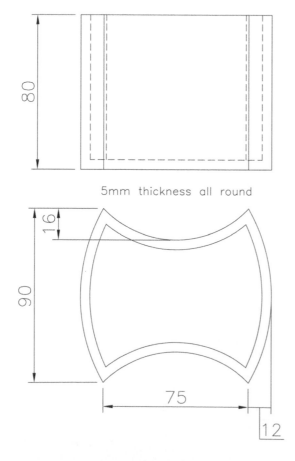

5mm thickness all round

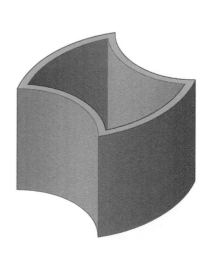

Exercise C18: bracket

Use the drawing information to create a part model of the bracket.

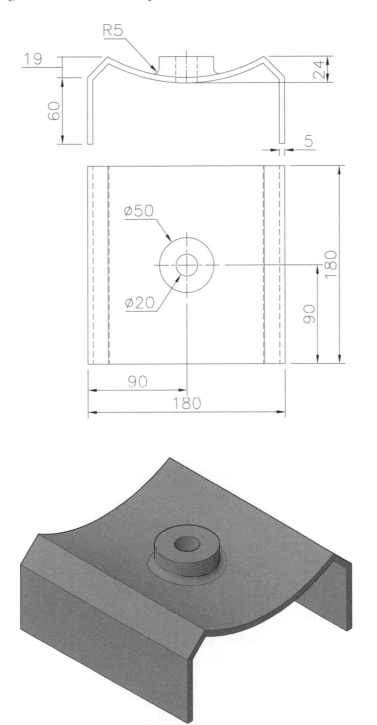

D   Work features

1   Work features are special construction features that are parametrically attached to parts.

2   They are used to help the user position and define new features in the model.

3   The following are three types of work feature:
   (a)   work axes;
   (b)   work points; and
   (c)   work planes.

4   Work features are used:
   (a)   to position a sketch for new features when a part face is not available;
   (b)   to assist with an intermediate position when other features need to be defined;
   (c)   to establish a plane or edge to assist with dimensions and constraints; and
   (d)   to provide an axis or point of rotation for revolved features and patterns.

5   In each of the D exercises, the process is:
   (a)   Start each exercise with a new standard metric (mm) .ipt file.
   (b)   Complete the sketch required.
   (c)   Ensure that the sketch is fully constrained.
   (d)   Add dimensions if necessary.
   (e)   Complete the part model.
   (f)   View in a suitable 3D viewpoint.
   (g)   Save your completed modified part model to your named folder (with a new name).
   (h)   Use discretion as appropriate.

Terminology

Work point
A feature that can be created at any time on an active part or in 3D space and can be:

1   used to assist with sketch geometry;

2   used to create work planes; and

3   created by various methods, including:
   (a)   at the intersection of three work planes or planer faces;
   (b)   at the intersection of two lines;
   (c)   at the vertex of a line or edge;
   (d)   on the midpoint of an edge; and
   (e)   at a point of intersection between a plane and a line.

Work axis

A feature that 'acts' like a construction line and can be:

1   used to create work planes;

2   used as an axis of rotation for polar patterns (arrays); and

3   created by various methods, including:
    (a) through a revolved face or feature;
    (b) through two endpoints, midpoints, intersections or work points;
    (c) along a linear edge; and
    (d) along a sketch line.

Work plane

1   Probably the most 'important' and most used of the work features.

2   A work plane is a feature that looks like a rectangular plane, and:
    (a) is used to sketch a feature (2D sketch) when no planer face is available;
    (b) can be edited and/or deleted like any other feature;
    (c) as many work planes as are needed can be created; and
    (d) can be created by various methods, including:
        •   by selecting any three points – endpoints, midpoints, intersections or work points;
        •   a curved face and a linear edge;
        •   two parallel planar faces or work planes;
        •   offset from a planar face;
        •   at an angle to a planar face; and
        •   tangent to a cylinder.

Exercise D1: creation of the three work features

1   To assist with the work feature creation exercises, refer to the two First Angle views below.

2   Create a part model using the detail as given, then save the completed model.

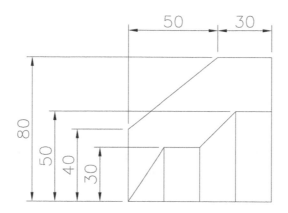

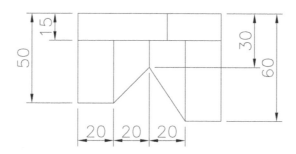

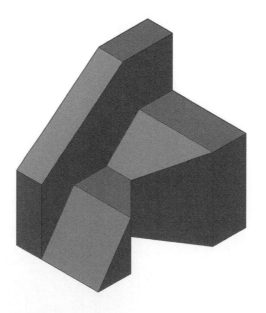

Exercise D1A: *creating work points*

Five work points to be created, on the vertices, as displayed.

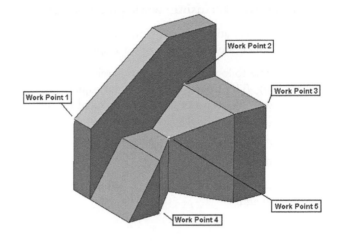

Exercise D1B: *creating work axes*

Create three work axes at the 'centre' of the three sloped faces indicated below.

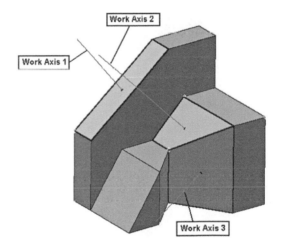

Exercise D1C: *creating work planes*

Create the following four work planes:

1 Through work points 1, 2 and 3.

2 Through work points 1, 4 and 5.

3 Offset 50mm from Face A.

4 Offset −100mm from Face B.

This completes the D1 exercise.

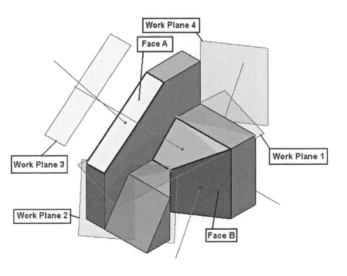

Exercise D2: using work planes to create a 'square to round'

1   A fabrication company produces funnels (or square to round) and has several models, three of which are displayed in First Angle projection below.

2   The following information is relevant:
   (a)   The funnel base is a 50mm square with height 40mm.
   (b)   The top is a Ø16mm circle and can be in three different positions.
   (c)   The funnel spout has a height of 30mm and ends with a 6mm diameter circle.

3   Create a part model of each funnel with a wall thickness of 0.5mm (or a thickness to suit the model).

4   Emboss the funnel type (i.e. letter A, B) on a 'side' of your choice, selecting your own embossed 'height'.

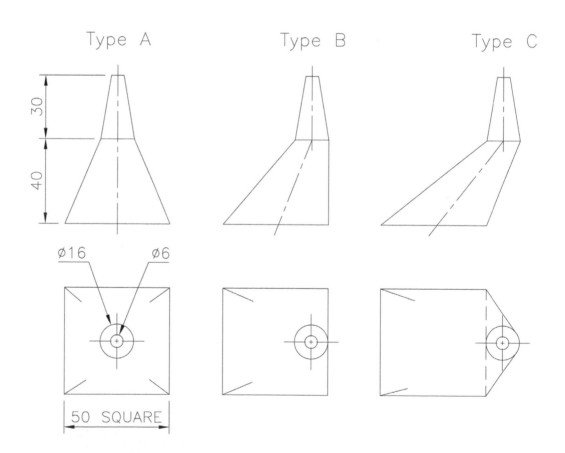

Funnel type A

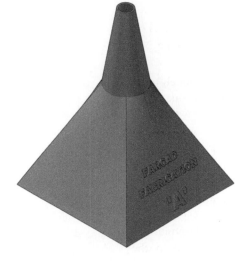

Funnel type B

Funnel type C

Exercise D3: funnel type D

1   The three funnels created in Exercise D2
    are termed by the company as 'in-line
    spouts'.

2   The company also customises funnels to
    customer specifications, and it is one of
    these 'specialised' funnels that has to be
    created.

3   The front and top views for this
    specialised funnel are displayed right, and
    the basic parameters are as before:
    (a) The funnel base is a 50mm square with
         height 40mm.
    (b) The top is a ⌀16mm circle positioned
         at a top vertex, as displayed in the
         diagram.
    (c) The funnel spout has a height of 30mm
         and ends with a 6mm diameter circle,
         but is offset 20mm, as displayed.

4   Create funnel type D with suitable text
    embossed on a side of your choice.

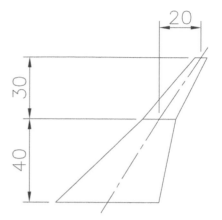

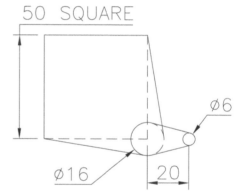

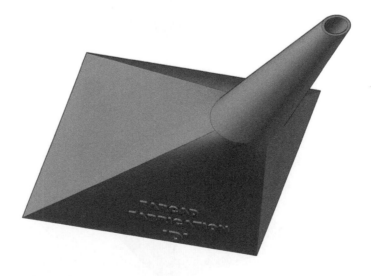

Exercise D4: work planes and circular components

1   Part of a pipe arrangement is displayed in three First Angle projection views below.

2   Use the information given to create a part model of the arrangement, adding suitable embossed text to any circular surface (remember discretion).

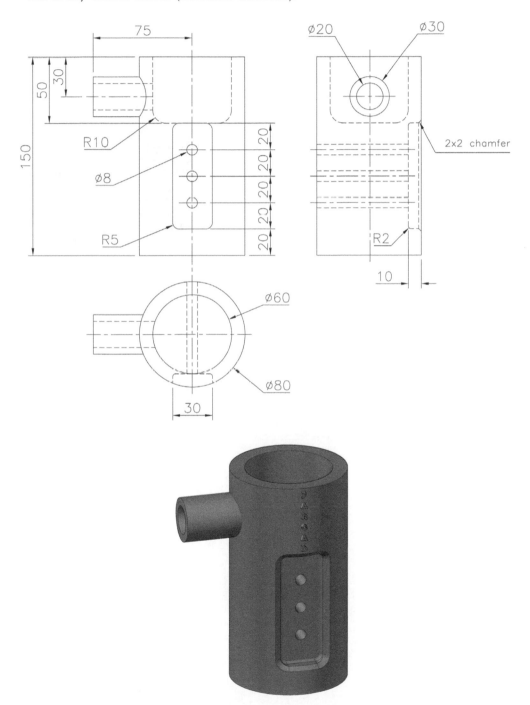

Exercise D5: pipe branch

1   Create the pipe branch using the information given in the two drawing views.

2   The pipes have a thickness of 10mm and are chamfered 5 × 5 at the ends.

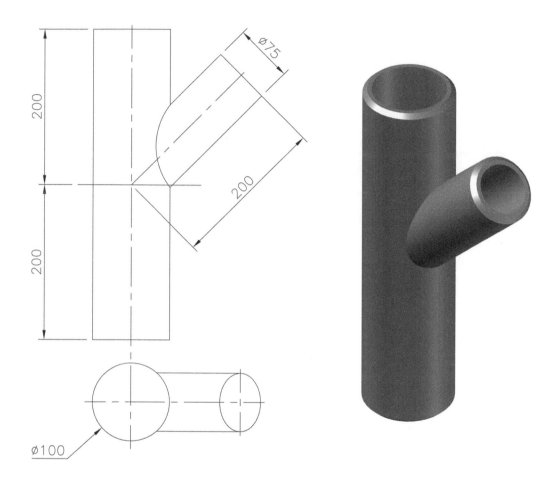

E Pattern features

1 Patterns allow the user to duplicate features multiple times:
   (a) for a set distance; and
   (b) for a set angle between the features.

2 For AutoCAD users, patterns are similar to arrays.

3 There are various 'types' of pattern available with Inventor®, including:
   (a) rectangular patterns;
   (b) circular patterns;
   (c) pattern along a path; and
   (d) mirror patterns.

4 With the E exercises, the procedure is the same:
   (a) Start each exercise:
      • with a new standard metric (mm) .ipt file; or
      • open a previously created and saved model.
   (b) Complete the new model using the drawing information given.
   (c) Ensure any modified (or new) sketch is still fully constrained.
   (d) Add dimensions if necessary.
   (e) Extrude or revolve the constrained sketch as appropriate.
   (f) View in a suitable 3D viewpoint.
   (g) Save your completed modified part model to your named folder (with a suitable name).
   (h) As usual, use your discretion as appropriate.

Terminology

Rectangular
The user selects a feature and specifies:

1 the number of repetitions (the count); and

2 the spacing for the rows and columns.

Circular
The user selects a feature and specifies:

1 the number of repetitions (the count);

2 the axis of rotation; and

3 the angle for the repetitions.

Path

The user selects a feature and specifies:

1   the number of repetitions; and

2   the created path.

Mirror

The user selects a feature and specifies the plane of symmetry.

Occurrence

All repeated features in a pattern are termed occurrences, and any individual occurrence can be suppressed.

Exercise E1: drain plate cover

1   Open the previously created part model of the drain plate cover – Exercise C14.

2   Modify the hole to 3mm diameter.

3   Create the rectangular pattern using the dimensions displayed, then save the model.

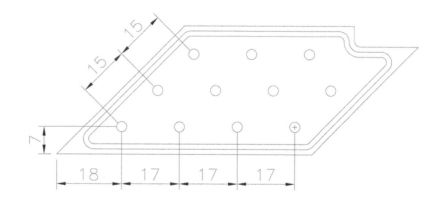

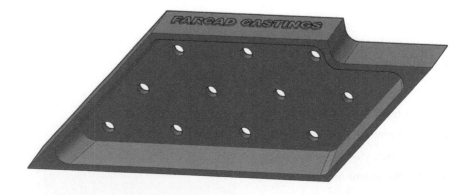

Exercise E2: connector

1   Open the previously saved part model of the connector – Exercise C15.

2   Create the circular pattern displayed.

3   Save the completed model.

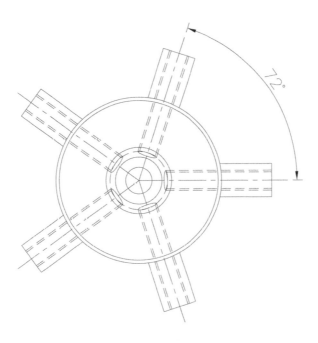

Exercise E3: plate cover

1   Use the drawing information to create the plate cover, which is 15mm thick.

2   The plate has to have an 8 × 8mm square hole pattern positioned as shown and a circular hole of diameter 30mm positioned at 'the plate centre'.

3   Create the part model with all features displayed and save the completed model.

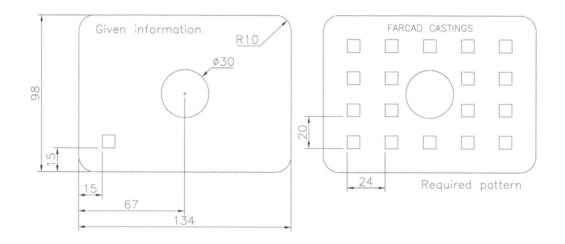

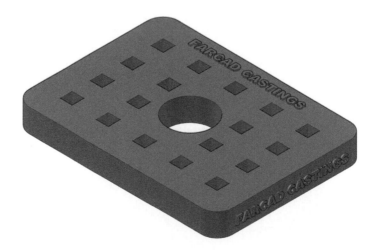

Exercise E4: flange ring

1   Using the information, create a flange ring of thickness 20mm.

2   The ring has a counter-bored hole with dimensions as displayed.

3   Complete the part model using the circular pattern tool (is a work axis needed?).

4   Save the completed model.

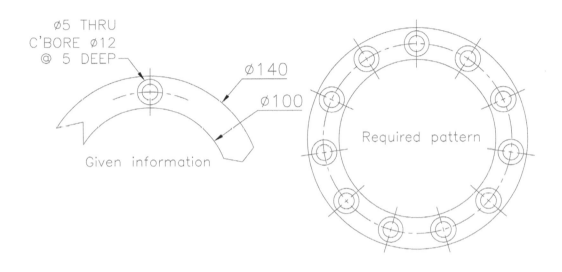

Exercise E5: manhole cover

1   A manhole cover has a diameter of 120mm and is 15mm thick.

2   The cover has:
   (a) a 5mm stud (4mm height with 1mm fillet around top) circular pattern of 27 items; and
   (b) a 5mm hole rectangular 'type' pattern, the hole distances being 10mm, both horizontally and vertically.

3   Use this basic information displayed to create the complete manhole cover.

4   Save the completed model.

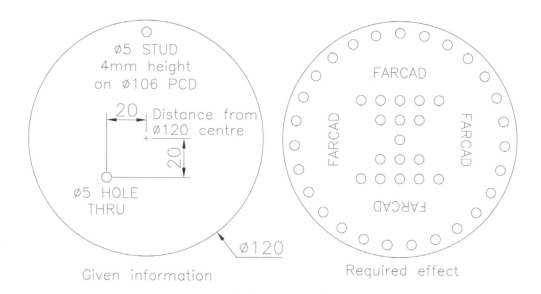

Given information   Required effect

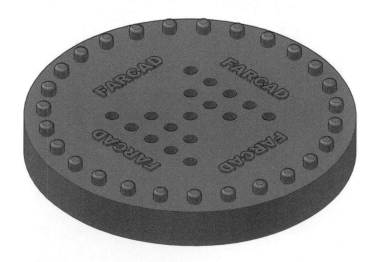

Exercise E6: film reel

Using the dimensions of one-quarter of a symmetrical film reel, create the complete part model and save. The reel is 'flat-faced' with a thickness of 22mm.

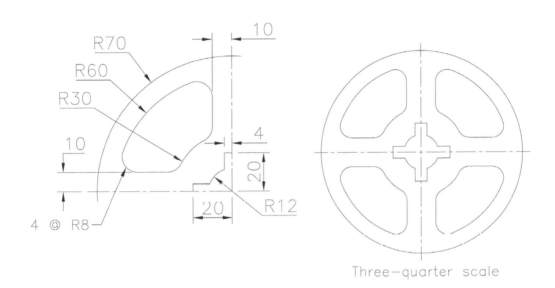

Three—quarter scale

Exercise E7: double loop hanger

Using the drawing information given, create a part model of the 60mm 'wide' double loop hanger.

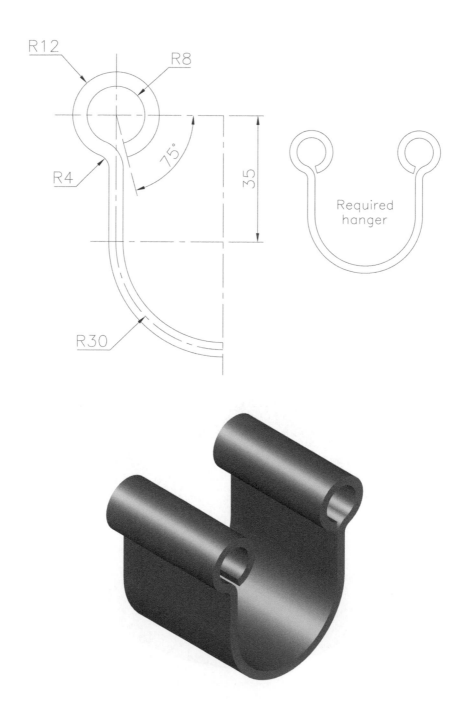

Exercise E8: tree with decorations

Open the saved Exercise A15 revolved part model of the tree and add the decorations using the information below. There has been no modification to the original tree outline.

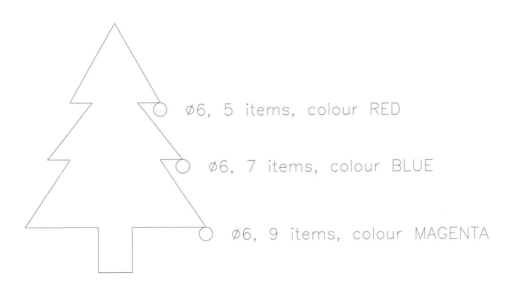

ø6, 5 items, colour RED

ø6, 7 items, colour BLUE

ø6, 9 items, colour MAGENTA

Exercise E9: gear wheel

1   A gear wheel has eight teeth and is 80mm wide.

2   The drawing below details the information required to construct one of the teeth.

3   Use this information to create the complete gear wheel.

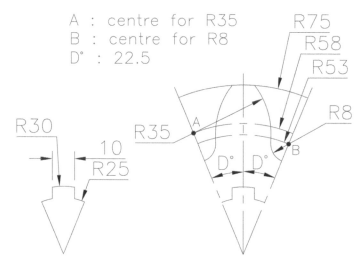

A : centre for R35
B : centre for R8
D° : 22.5

Construction details

Complete gear wheel profile

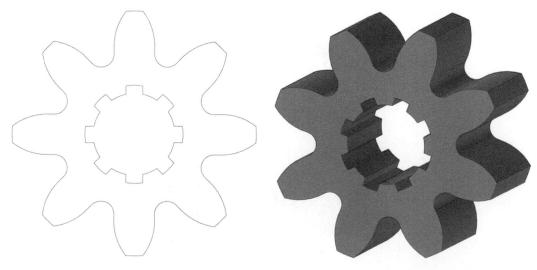

Exercise E10: clock face

Using the simple drawing below, create a part model of a clock face, using your design ability for the numeric display. The clock face is 20mm 'thick'.

Exercise E11: extruded logo

1   The CAD HERITAGE foundation has asked a plastic moulding company to design a suitable logo for their new premises.

2   The company has decided that the letters C and H could be suitable, and has produced its initial 'run' of the design.

3   Using the basic information in the drawing below, create a part model of this moulding design, using your discretion as appropriate (e.g. colour, layout, etc.).

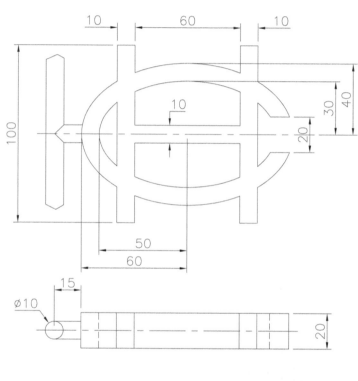

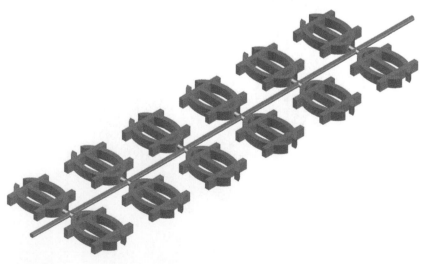

Exercise E12: car wheel design

1   Create some concept car wheel design.

2   The only restriction is that the overall diameter is 100mm – the rest is up to you.

Exercise E13: fisherman's disc

1   A fisherman has a disc (∅150mm and 5mm thick) with the outline of a fish 'embossed' for 3mm on the face side.

2   The fish shape is as shown in the diagram below.

3   Create a part model of the disc using your own 'design' for the layout.

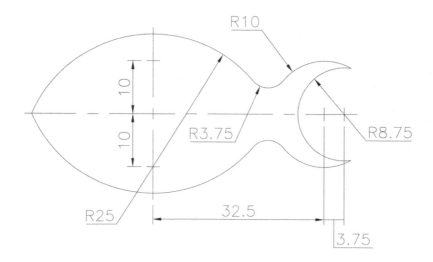

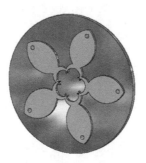

Exercise E14: spacer

1 Create a part model of the spacer using the drawing information (and your discretion) given below.

2 The 'central shape' is half-extruded for 5mm and half-cut for 5mm.

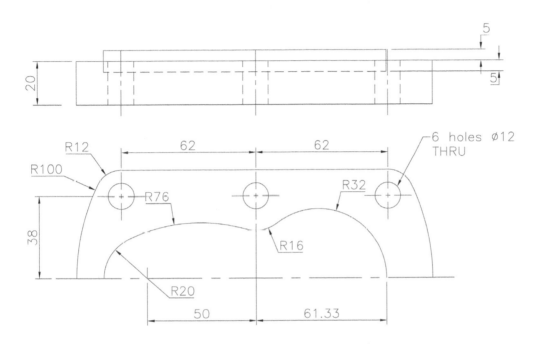

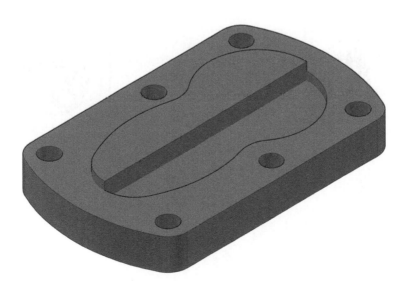

Exercise E15: picture frames

1   Two picture frames have to be created, based on the drawing data below.

2   Both frames have a 'thickness' of 4mm and have a 3mm recess in the front face.

3   The frame A recess has been detailed, but the frame B recess is for you to design.

4   If possible, import a suitable image for the frames.

FRAME A

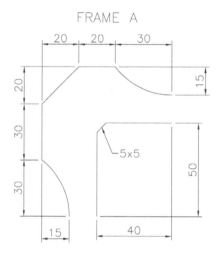

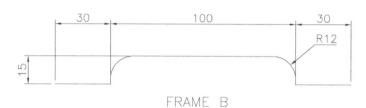

FRAME B

Exercise E16: sporting ornament

1   A sporting ornament has been designed using the drawing information given.

2   The main 'arrow part' of the ornament is inclined at 45 degrees to the horizontal.

3   The circular base of the ornament is a torus shape of 120mm mean diameter with a 20mm (or suitable diameter) tube diameter.

4   Create a part model of the ornament using your own pattern design.

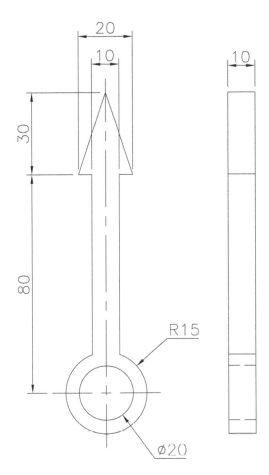

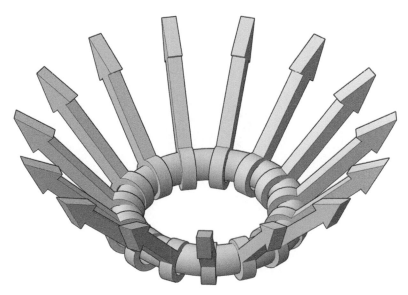

Exercise E17: stool

1   You have to create a part model of a stool with four legs and a top.

2   The only information given is the very basic sketch below.

3   The final shape of the leg is for you to design.

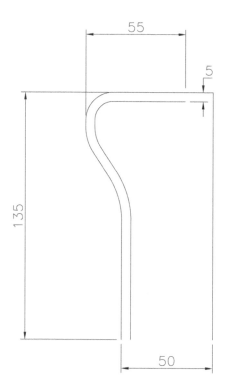

Exercise E18: leaf

1 Use the information below to create a part model of the leaf.

2 When complete, create a pattern of your own.

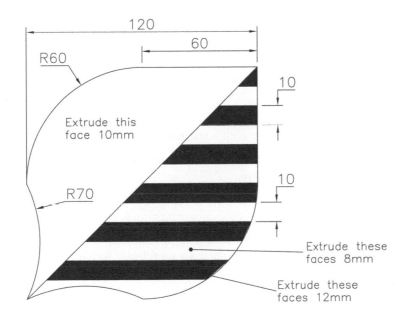

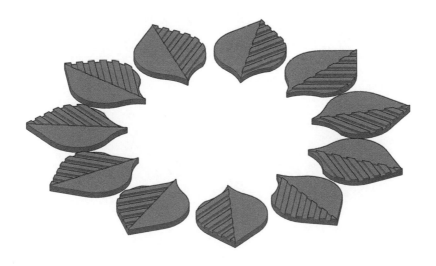

F   Allsorts

1   Patterns allow the user to duplicate features multiple times:
    (a)   for a set distance; and
    (b)   for a set angle between the features.

2   This chapter will present the user with a series of varied (I hope) exercises.

3   The exercises could be simple extrusions and revolutions, but may also require the creation of work planes and the use of patterns.

4   There is no order for the exercises as I have added them as they came to mind.

5   In each exercise, the procedure is the same:
    (a)   Start each exercise with a new standard metric (mm) .ipt file.
    (b)   Use the information given to complete an appropriate sketch.
    (c)   Ensure that the sketch is still fully constrained.
    (d)   Add dimensions if necessary.
    (e)   Create the Inventor® part model.
    (f)   View in a suitable 3D viewpoint.
    (g)   Save your completed part model to your named folder. It may be used for the I exercises.
    (h)   Use discretion as appropriate – very important.

Exercise F1: step extrusion

1   Use the two given views with dimensions to create an extruded step effect.

2   Save the completed model.

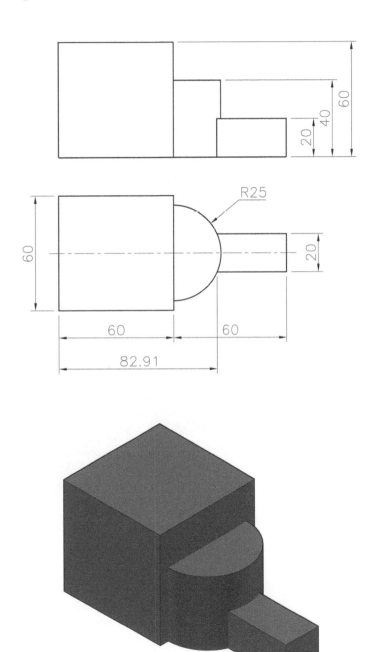

Exercise F2: mallet

Use the profile details below to create a revolved model of the mallet.

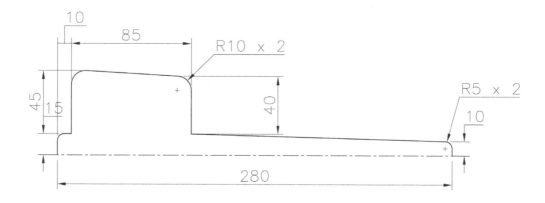

Exercise F3: ducting corner

1 Use the information to construct a SWEPT model along the given path.

2 When created, shell for a thickness of 2mm.

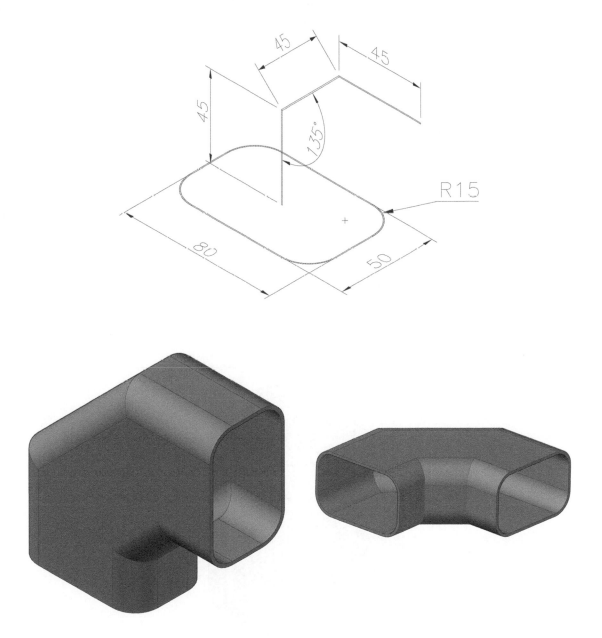

Exercise F4: half a jar

Using the outline details, create a REVOLVED 'half-jar' part model.

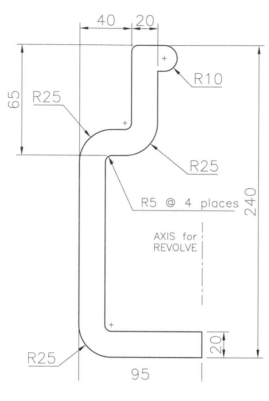

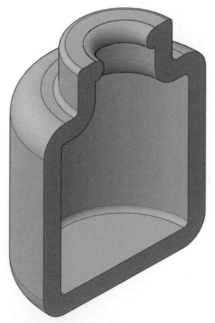

Exercise F5: swept flange

1   Create the I beam outline using the suggested projection point.

2   Create the sweep path using the sketch details given.

3   Construct a swept part model of the I beam.

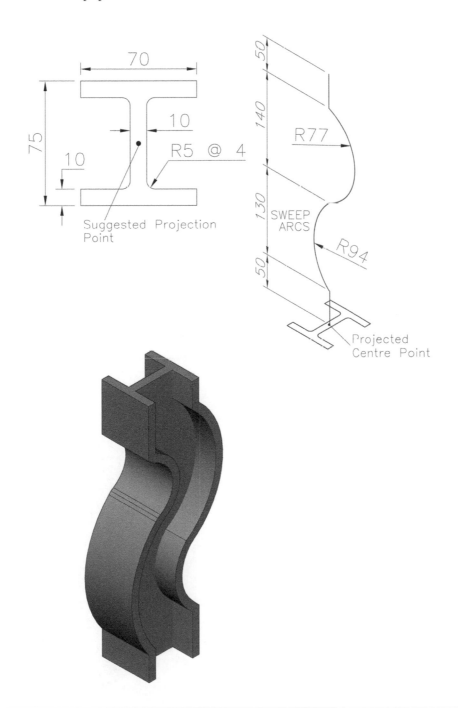

Exercise F6: loft trophies

1   Two designs based on the loft tool.

2   Use the information given to design a 'lofted trophy'. The circle centres are all on the 'same vertical centre line' and are 30mm 'offset' from each other.

3   Another 'lofted trophy' but this time no dimensions are given. Use the layout as a starter with your own sizes.

4   Add any refinements and save each model.

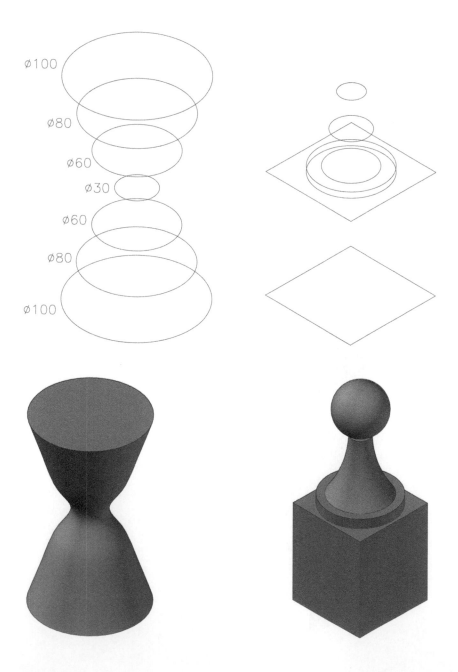

Exercise F7: bungalow layout

1   Create a part model of the bungalow using the plan details given.

2   The bungalow height is 100mm and the wall thickness is 5mm.

3   Add windows and doors to your own specification and any other refinements.

4   Save the model when complete.

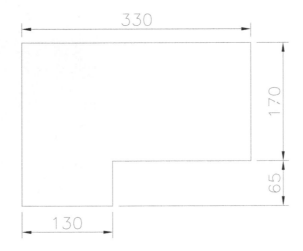

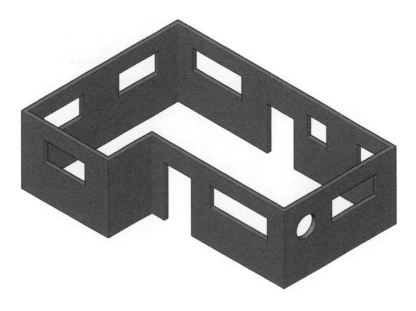

Exercise F8: wine glass

1   Use the dimensioned profile to create a part model of the wine glass.

2   The outline shape is not as straightforward as you may think.

3   Remember to save the model.

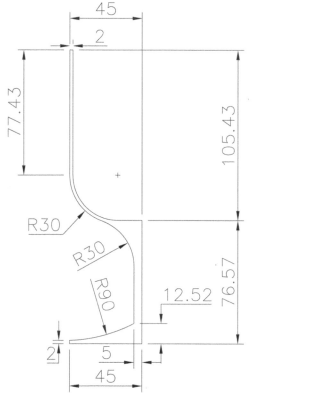

Exercise F9: moulding 1

1   Create the path and profile on two work planes, then create the swept moulding.

2   Save for the next exercise.

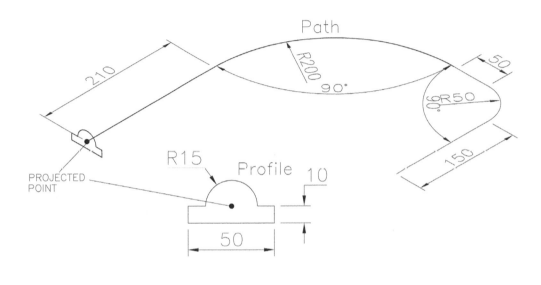

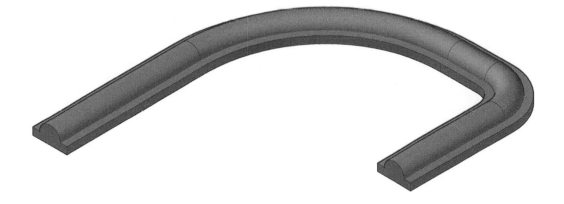

Exercise F10: moulding 2

Use the path and the basic profile from Exercise F9, and:

1   Rotate the moulding profile 180 degrees about the projected centre point.

2   Extrude the moulding profile for 10mm and create a 10mm square hole, centred on the projection point.

3   Create a pattern using the new rotated and extruded moulding with the 10mm hole for:
(a) 20 items;
(b) along the original path; and
(c) the moulding should start and end at the 10mm swept path.

4   Refine your layout, then save.

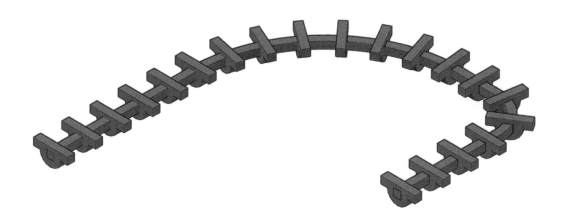

Exercise F11: mace

Using the 'relatively easy' profile below, create a three-quarter revolved part model of the mace, using your discretion as usual, and then save when the model is complete.

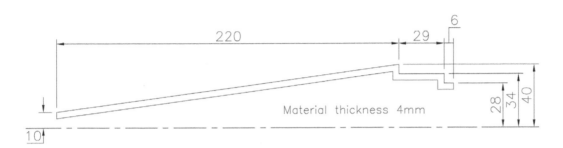

Exercise F12: pyramid

1   An 85mm pyramid rests on an 80mm square base with a 5mm square top.

2   The pyramid has a circular hole and a square hole as detail in the dimensioned drawing.

3   Create a part model of the pyramid and save when complete.

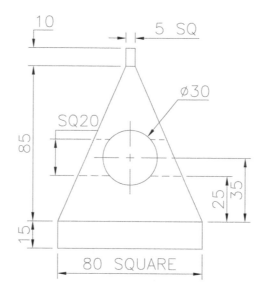

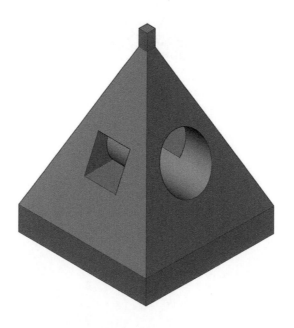

Exercise F13: another pinion gear

1 A pinion gear has to be created using the information given in the detailed drawing with the following data:
   (a) Outside tooth diameter: 140.
   (b) Root diameter: 95.
   (c) Shaft diameter: 60.
   (d) Crest radius: 2.
   (e) Root radius: 2.
   (f) Number of teeth: 12.
   (g) Width of gear: 20.

2 Using the given data, create the pinion gear then save the completed model.

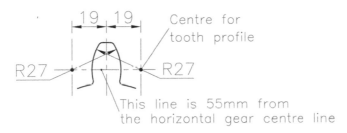

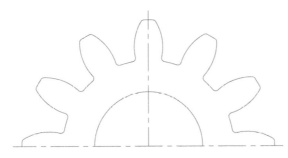

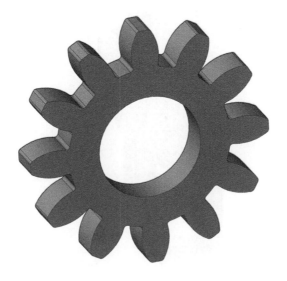

Exercise F14: wall clock

1   A wall clock consists of three parts: a body, a face and a dial.

2   The information to create each part of the clock is given in the detailed drawings.

3   Using the information, create a part model of the clock (not an assembly).

4   Note that the centre point of the face and dial is 120mm from the lower 'apex' of the body.

5   As an additional extra, add suitable parts to the face – I decided on an extruded circle as it was the easiest option, but you can design your own.

6   *Note:*
   (a) no hands have been included; and
   (b) we will use this exercise with assemblies.

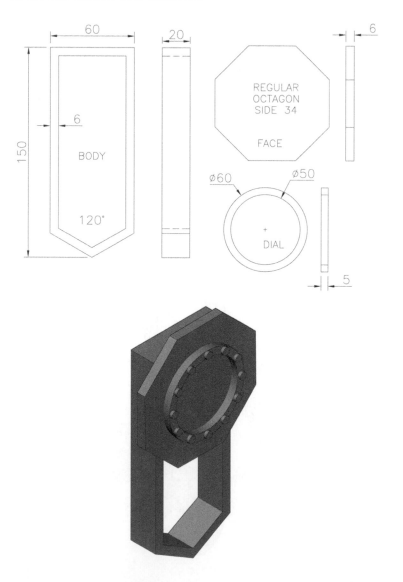

Exercise F15: pot stands

A designer in a ceramics company was creating pot stands. One of the stands was based on a square grid concept, the other on an X shape. Using the drawings supplied, create part models of the two pot stands, both of which are 15mm in height. Add any refinements as required.

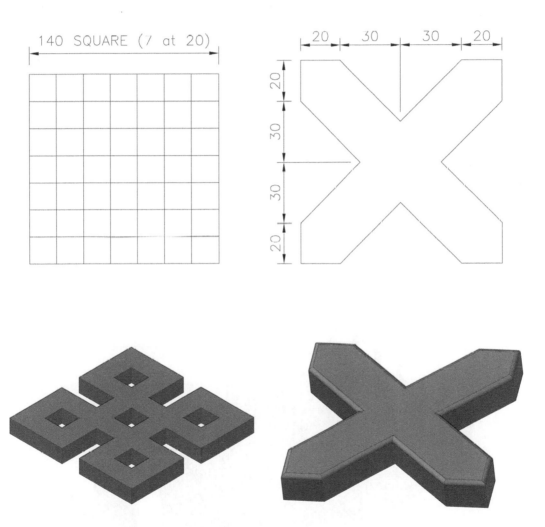

Exercise F16: model factory layout

Before constructing the building, the construction company built a model of the factory layout, details of which are displayed below in plan view. You have to create a part model of this layout using the drawing information given and your imagination, especially when creating the I beam.

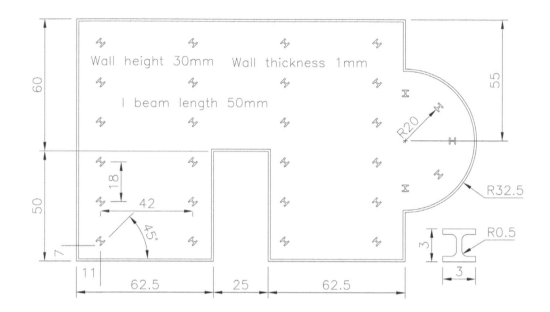

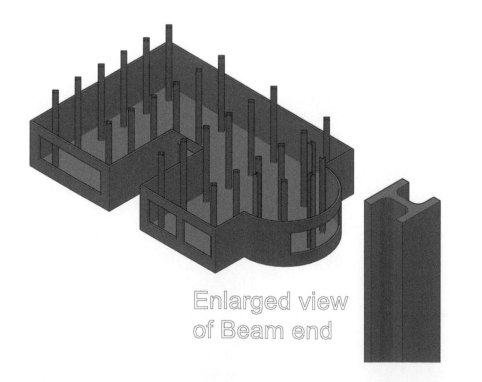

Enlarged view
of Beam end

Exercise F17: plastic spring

Use the dimensioned drawing below to create a part model of the plastic spring that is 50mm wide.

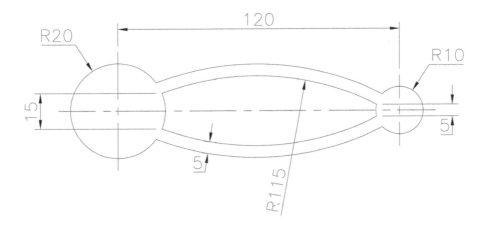

Exercise F18: plastic bowl

Use the drawing information given to create a part model of the plastic bowl.

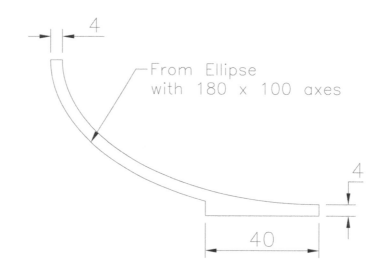

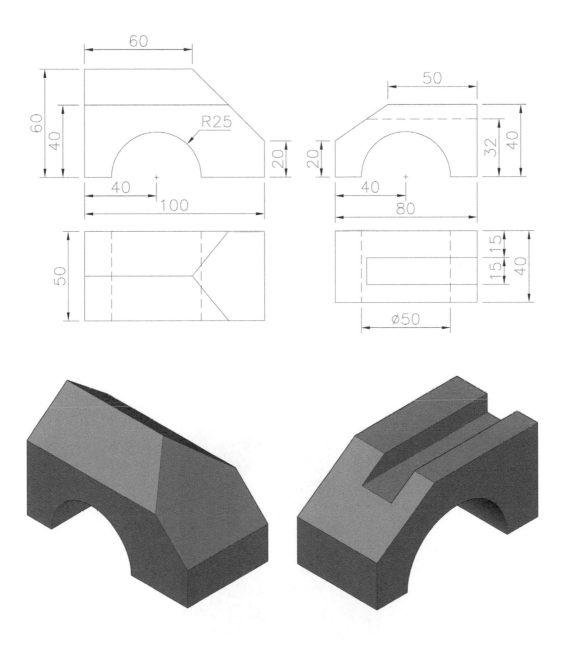

Exercise F19: supports

Create two part models of supports using the drawing information given.

Exercise F20: spout

Create a part model of the component using the two First Angle projection views given.

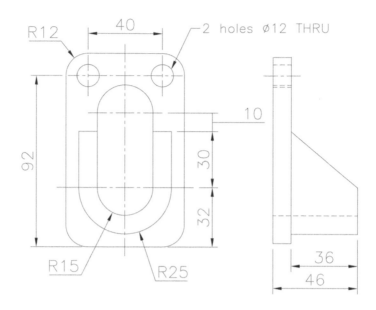

Exercise F21: fan motor housing

Using the two First Angle projection views given, create a part model of the fan housing, which is an interesting model to complete.

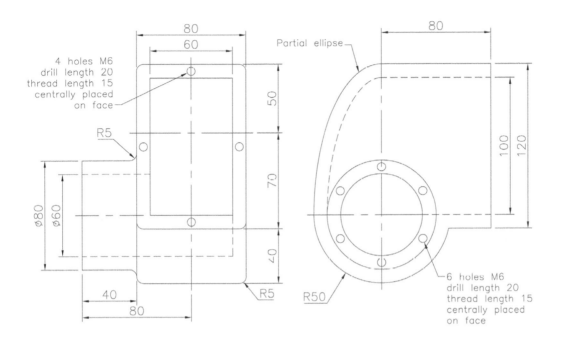

Exercise F22: cones

Create three cone part models using the drawing information given.

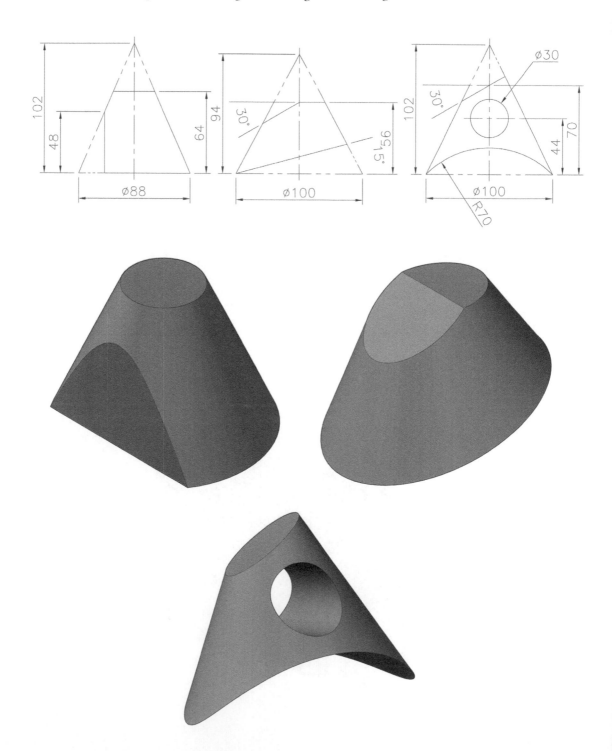

Exercise F23: paper clip

Create a part model of the paper clip by sweeping a 2.5mm diameter circle along the path displayed (the path is interesting to create).

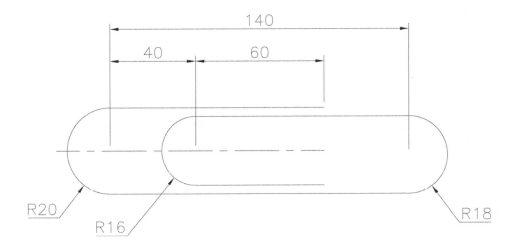

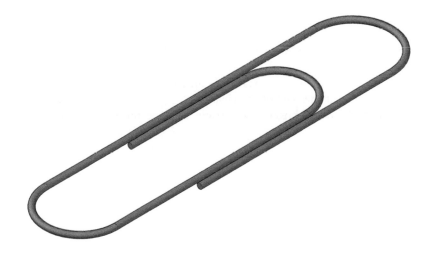

G Engineering

1   The models to be created in this chapter are 'basically' engineering-based.

2   The information given to create the models will be in the form of 2D orthographic views (in First or Third Angle projection) of the model, so you will need to 'read' the drawings to obtain the required (and correct) data.

3   In each exercise, the procedure is the same:
    (a) Start each exercise with a new standard metric (mm) .ipt file.
    (b) Use the information given to complete an appropriate sketch.
    (c) Ensure that the sketch is still fully constrained.
    (d) Add dimensions if necessary.
    (e) Create the Inventor® part model.
    (f) View in a suitable 3D viewpoint.
    (g) Save your completed part model to your named folder.
    (h) Use discretion as appropriate – very important.

4   *Note*:
    (a) As with exercises in other chapters, there is no order to the difficulty of the models to be created.
    (b) They are placed in the order I created them, so some at the end of the chapter may be 'easier' than some at the start of the chapter.
    (c) Adding materials and embossing can greatly enhance the model appearance, but this is your decision.

Exercise G1: parallel-sided cover plate

1   Create a part model of the component from the given views.

2   The central ⌀50mm 'hole' is 55 deep.

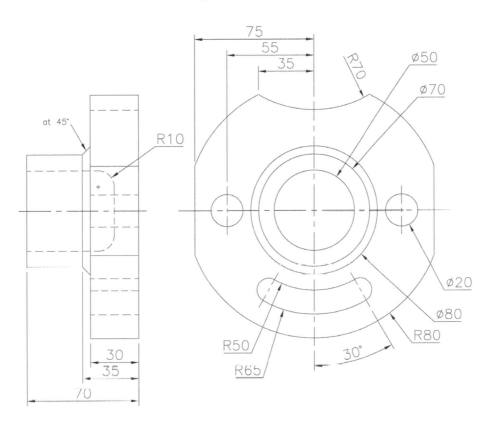

Exercise G2: protected bearing housing

Create a part model of the bearing housing from the two given views, noting:

1   the fillet radii are 5mm and 10mm;

2   the wall thickness is 10mm or 20mm; and

3   saving when complete.

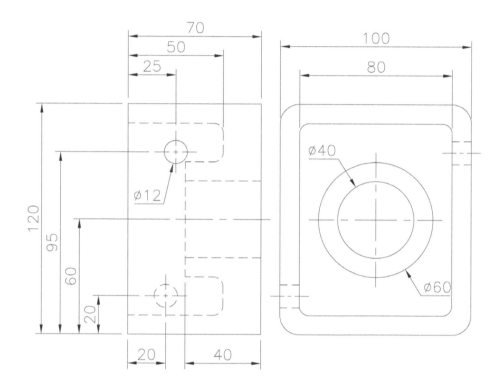

Exercise G3: steam expansion box

1 Details and sizes of the steam expansion box are given below.

2 Use this information to create a part model of the component with the final display being at your discretion.

3 Note that the wall thickness is 10mm.

4 Save the completed model.

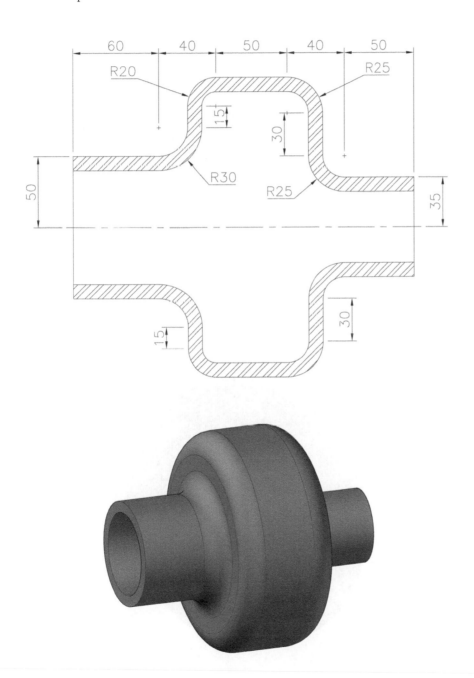

Exercise G4: gasket cover

1   The details of the gasket cover are displayed below.

2   Use this information to create a part model of the part and display the 'full' completed model, then save.

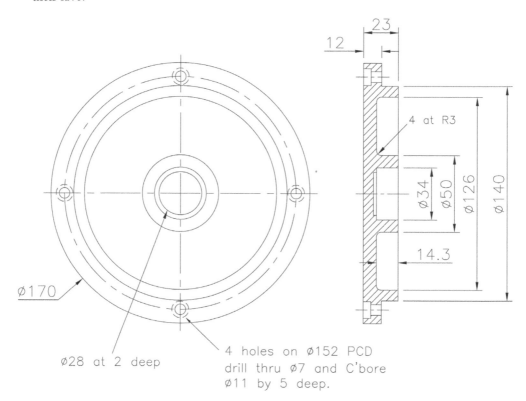

Ø170

Ø28 at 2 deep

4 holes on Ø152 PCD
drill thru Ø7 and C'bore
Ø11 by 5 deep.

23

12

4 at R3

Ø34
Ø50
Ø126
Ø140

14.3

Exercise G5: screwdriver

A slightly different exercise this time:

1 Create a part model of the screwdriver using the information given in the drawing, using your discretion as appropriate.

2 This model 'combines' both the revolved and extruded part features.

3 The 'flat part' of the driver is 3mm thick and the point is 1mm thick.

4 Apply suitable materials before saving the completed model.

5 *Note*: my distance/dimension A was 21.541mm.

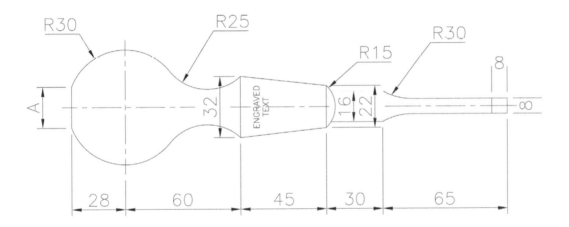

Exercise G6: light bulb

1 Create a model of the light bulb using the information given in the drawing and using your discretion as appropriate and save the completed model.

2 *Task*: with a bulb shell of 0.5mm, can you add simple elements 'inside'?

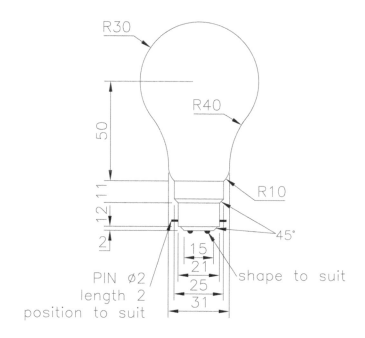

Exercise G7: bearing block

1   Two traditional First Angle orthographic views of the component are displayed.

2   Use these views to create an Inventor part model of the bearing block. *Note*: I have 'assumed' two chamfers at A, but without an 'end view' there are other options. There are also various options for the 'feature' labelled B.

3   Save the completed model.

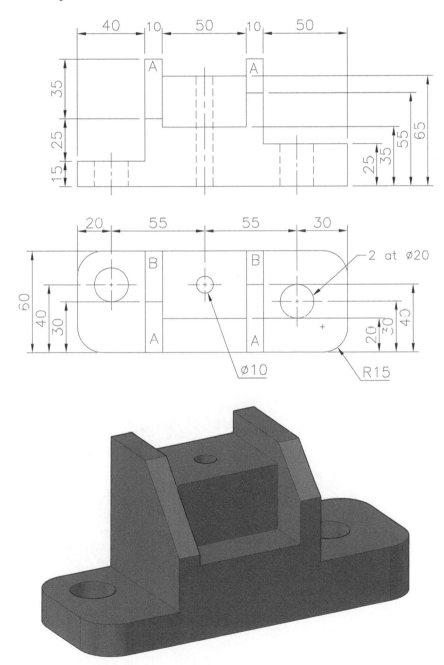

Exercise G8: positioning plate

Using the two given orthographic views, create a part model of the component and save when complete.

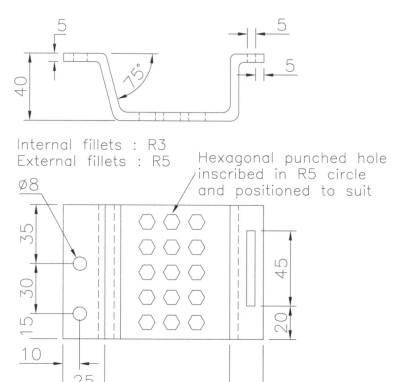

Internal fillets : R3
External fillets : R5

Hexagonal punched hole inscribed in R5 circle and positioned to suit

Exercise G9: bicycle spanner

Use the details in the views below to create a part model of the bicycle spanner. A fairly easy exercise. Save the completed model.

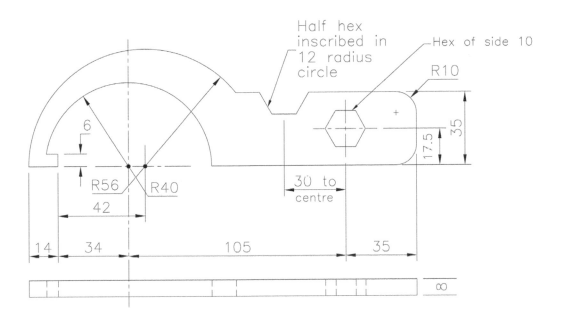

Half hex inscribed in 12 radius circle

Hex of side 10

R10

35

17.5

30 to centre

6

R56

R40

42

14

34

105

35

∞

Exercise G10: Swiss mechanism drive

From the two First Angle orthographic views, create a part model of the mechanism drive, and save the completed model.

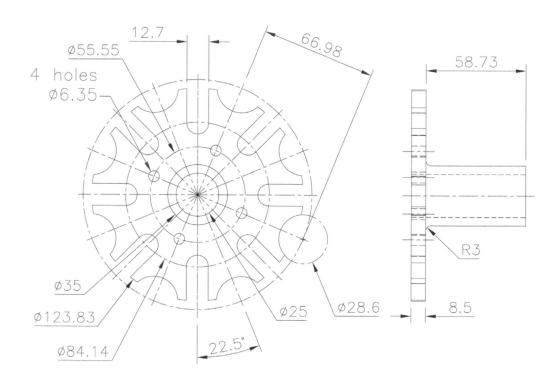

Exercise G11: vent cover plate

1   Using the given reference sizes, create a model of the vent cover plate.

2   The plate thickness is 5mm and the thickness of the 'vents' is 2mm.

3   Save the completed model.

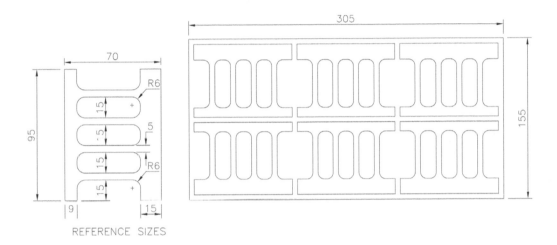

REFERENCE SIZES

Exercise G12: flat-bladed propeller

1   Refer to the dimensioned sketch and create a part model of a single propeller blade on a 'drive shaft'.

2   The blade thickness is 5mm and the shaft length is 30mm.

3   When the basic blade has been created, you have to create a two-, three- and four-bladed arrangement, and save as appropriate.

4   There is a second part to this exercise after the three arrangements have been completed.

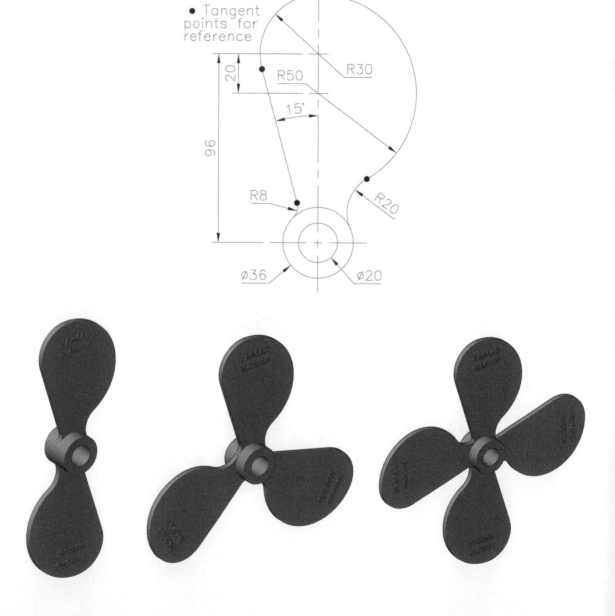

5 The original single blade has to be modified as displayed in the sketch right:
  (a) the basic blade shape is unchanged; and
  (b) the shaft is re-aligned and is now 25.16mm in length.

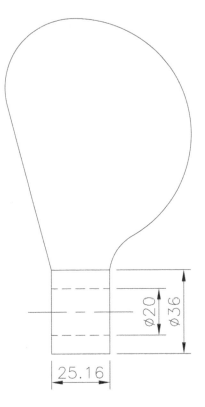

6 For this orientation, create a two-, three- and six-bladed propeller arrangement and save each one.

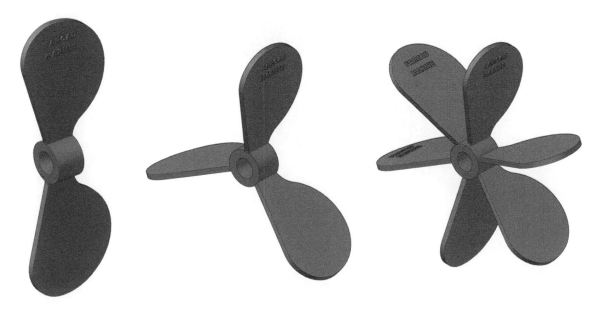

Exercise G13: rocker arm

1   Refer to the dimensioned sketch and create a part model of the rocker arm.

2   The rocker arm has a thickness of 5mm and the shaft projects 8mm either side of the arm and save when complete.

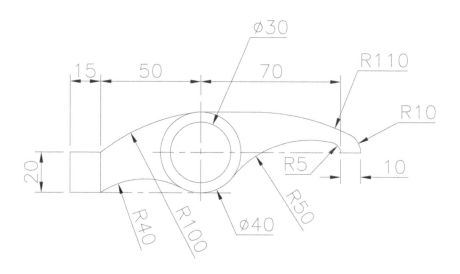

Exercise G14: backing plate

1   Use the information in the three First Angle orthographic views to create a part model of the backing plate (the side slots have a width of 10mm).

2   Save when complete and use discretion as usual.

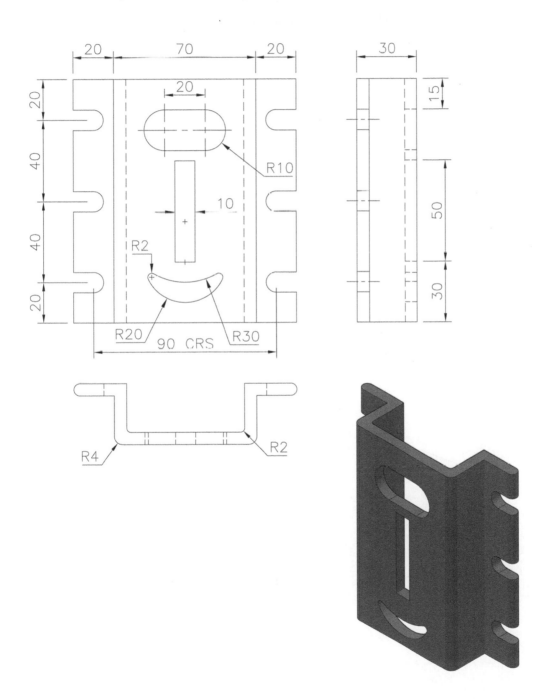

Exercise G15: desk tidy

1   Using the information displayed, create a part model of the desk tidy.

2   The front slot is to be your own design.

3   The horizontal surface at the top of the sloped end has a width of 10mm.

4   Save when complete and use discretion as usual.

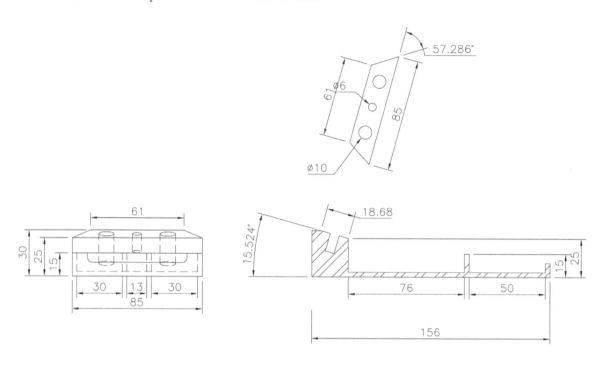

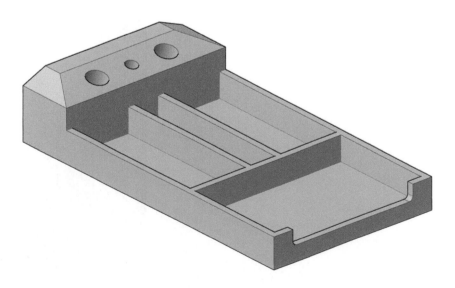

Exercise G16: bearing support

Create a part model of the bearing support using the drawing information given and save when complete.

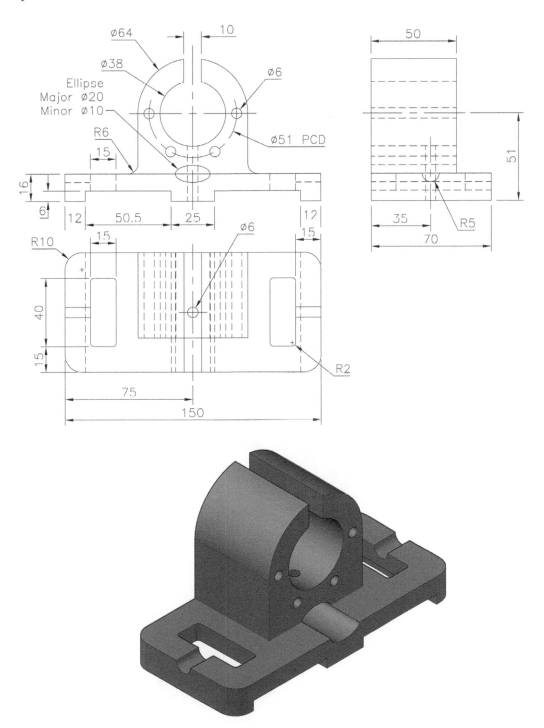

Exercise G17: table lamp

Using the drawing information displayed, create a part model of the table lamp and save when complete.

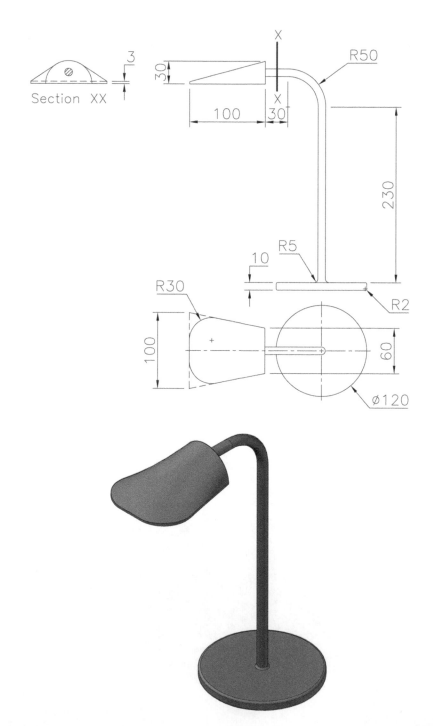

Exercise G18: micrometer

Use the drawing information displayed to create a part model of the micrometer. Save when complete and use discretion as usual.

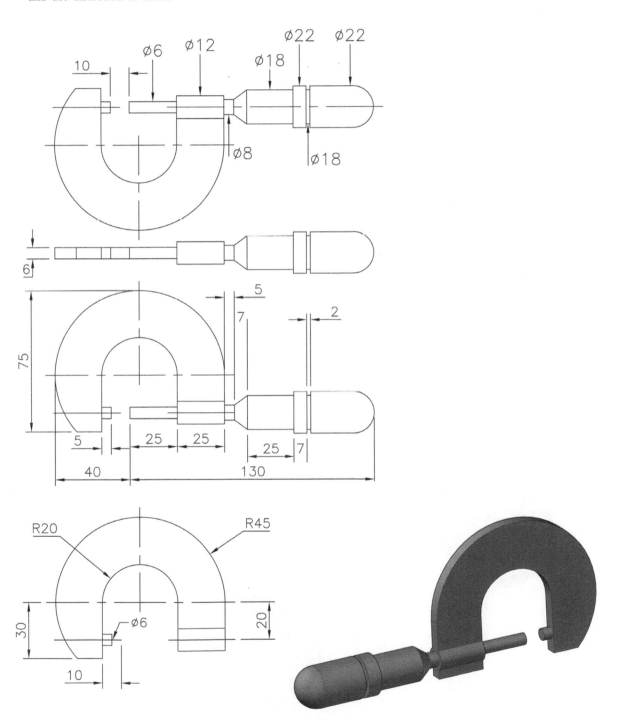

Exercise G19: adhesive tube

Use the drawing information displayed to create a part model of the tube and save when complete.

Question: Can you add a SHELL effect?

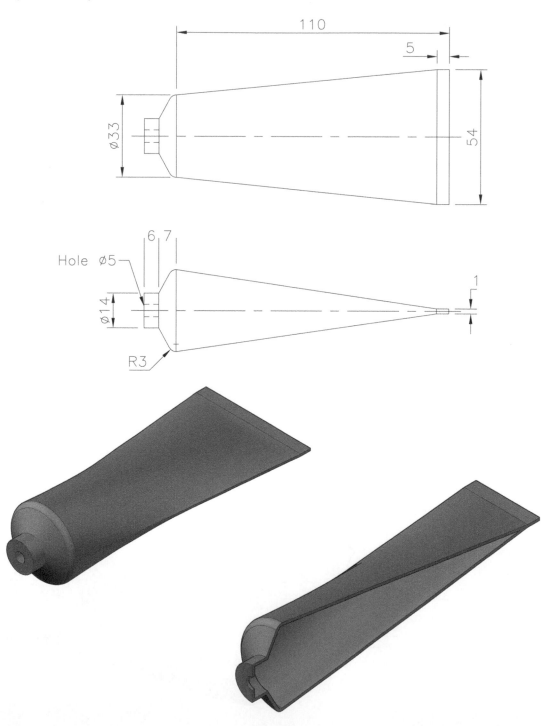

Exercise G20: brake shoe

Use the drawing information displayed to create a part model of the component and save.

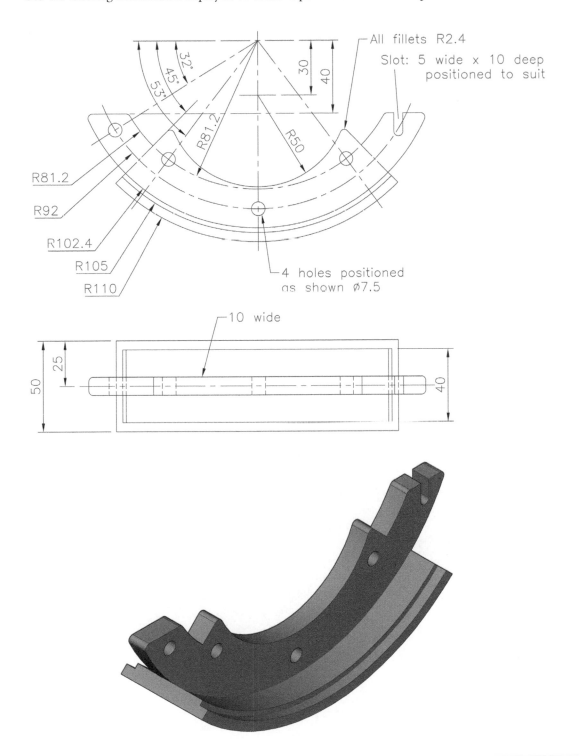

Exercise G21: support

Create a part model of the support using the three Third Angle views given.

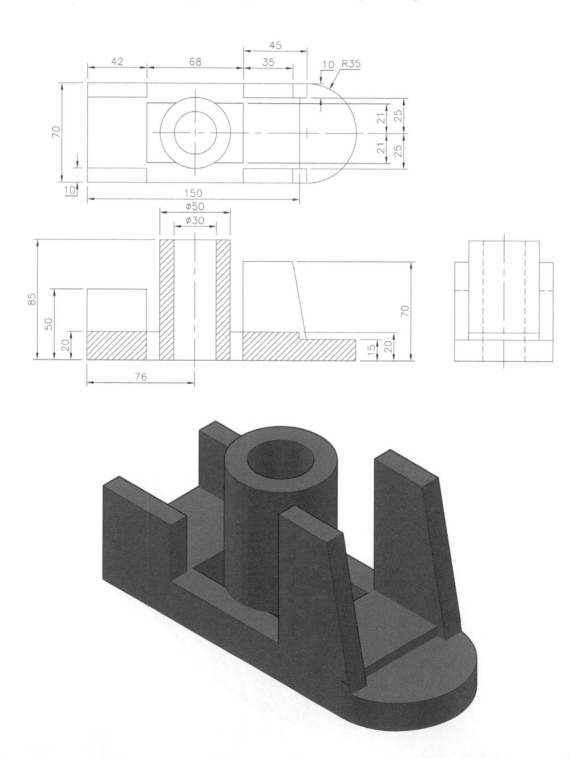

Exercise G22: dipstick

Using the drawing information below (path, outline), create a part model of the dipstick, adding any refinements of your choice.

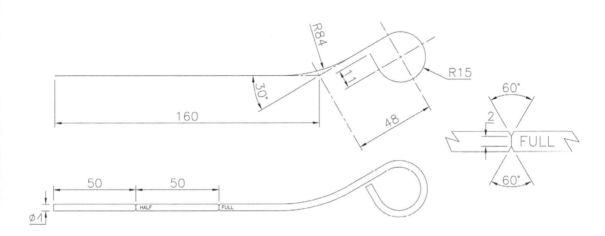

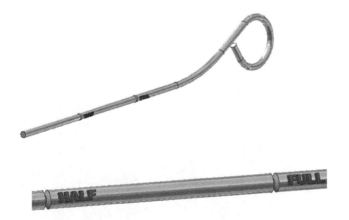

Exercise G23: fixture

Use the dimensioned solid model drawing to create a part model of the component.

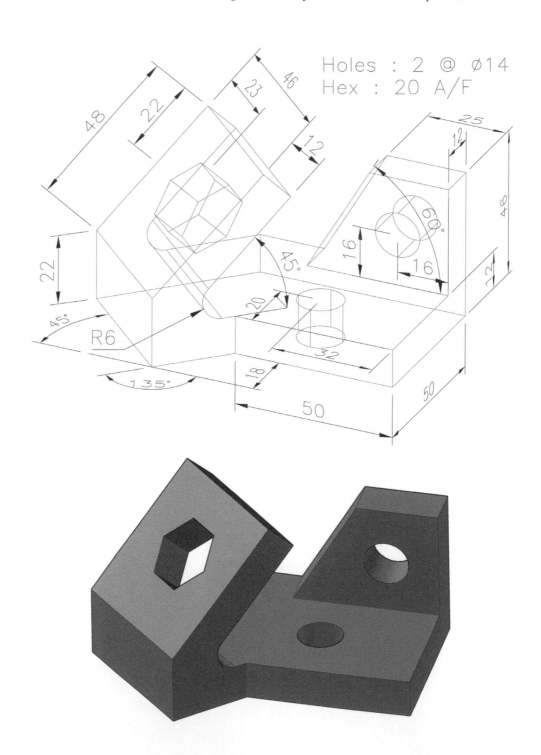

Holes : 2 @ ⌀14
Hex : 20 A/F

Exercise G24: ducting

Two ducting parts to be created with a material thickness of 1mm.

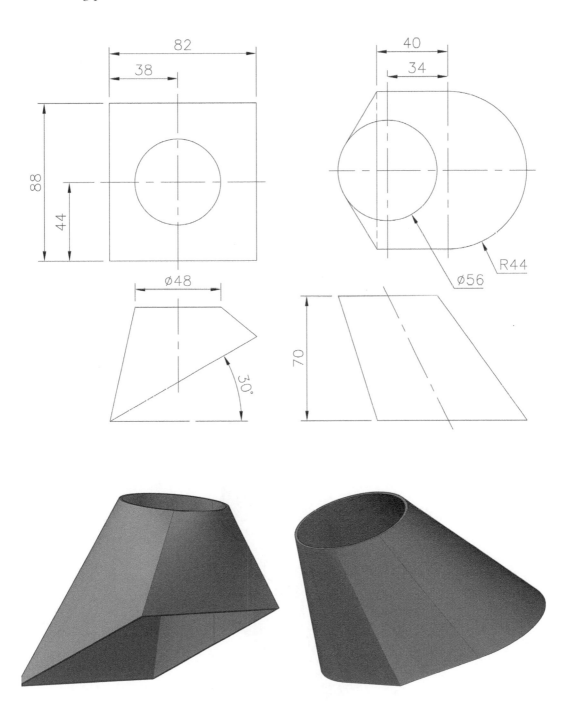

Exercise G25: bracket

Two Third Angle projection drawings to enable you to create the part model.

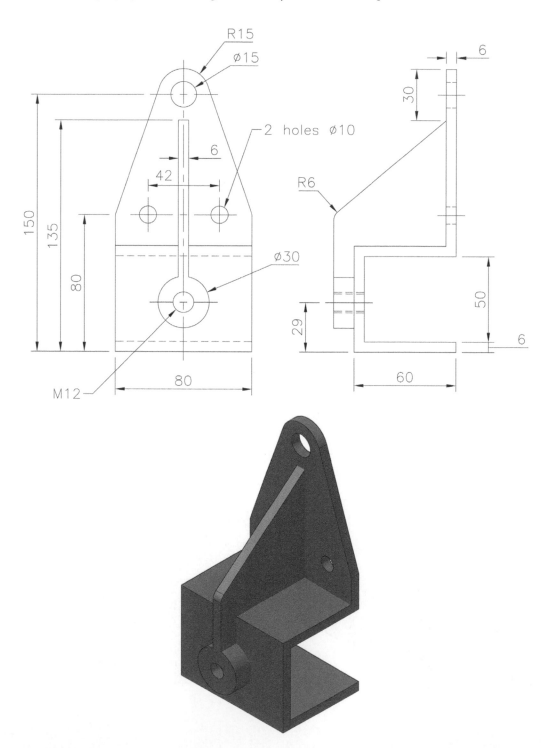

Exercise G26: spring

Create a 5mm diameter copper spring using the path outline displayed.

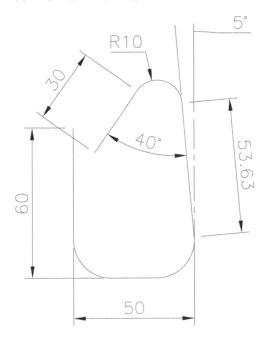

Exercise G27: support

Three First Angle views for the creation of the part model component.

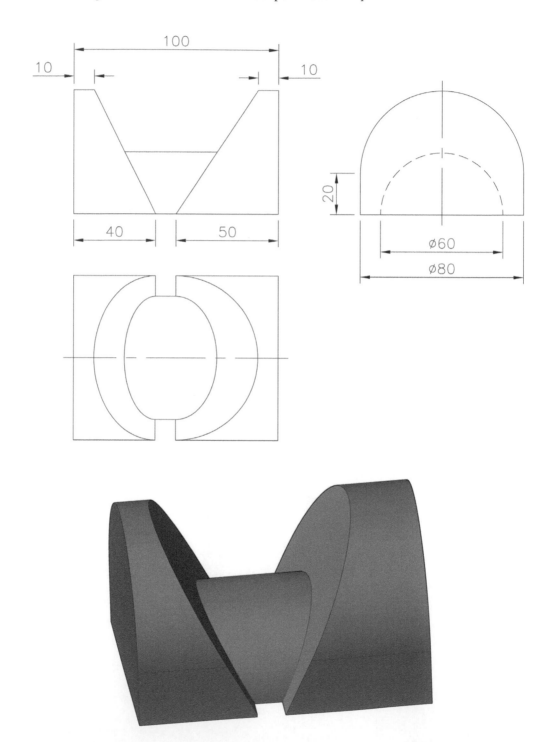

Exercise G28: tapered block

Three First Angle views for the creation of the part model component.

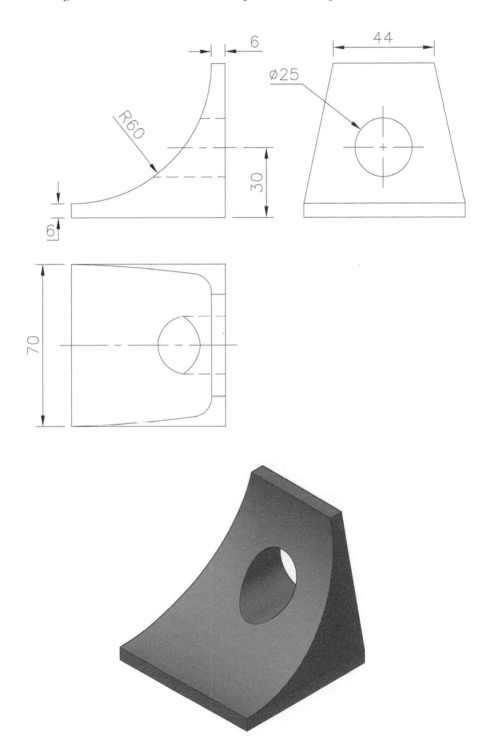

H   Assemblies

1   An assembly is placing (positioning) previously created part models into an *.iam file* and applying constraints between various components as appropriate.

2   There are various methods for creating assemblies, including:
   (a)   bottom-up, in which all of the components were created in individual .ipt files and then referenced into the assembly;
   (b)   top-down, in which the individual components are created 'in the Assembly file'; and
   (c)   combination, which uses both the bottom-up and top-down techniques.

3   *Note*:
   (a)   There is no correct or incorrect way to create an assembly.
   (b)   The various assembly models in the chapter exercises have been created using the bottom-up technique.
   (c)   The user can use any method to complete the assembly exercises.

4   In each assembly exercise, the procedure is the same:
   (a)   Create each component of the assembly as a separate part model using the standard metric (mm) .ipt file with discretion as usual.
   (b)   Save each part model with a suitable name.
   (c)   Open a new standard metric (mm) .iam file.
   (d)   Place the first part in the drawing area – the *GROUND component*.
   (e)   Insert (place) the other parts of the assembly as appropriate.
   (f)   Mate the placed components using suitable constraints and remember that some components may require two or three mating operations.
   (g)   View at a suitable 3D viewpoint.
   (h)   Add refinements of your choice (e.g. materials, embossing, logo, etc.).
   (i)   Save the assembly.
   (j)   The information required to create the individual parts for the assembly will be given as drawing views in either First or Third Angle projection.

5   *Suggestion*: give each part model in an assembly a different material/colour for a 'more pleasing' assembly.

6   Take a look at ASSY-H1.iam and ASSY-H1.idv on the companion website (www.routledge.com/cw/mcfarlane), in the folder of sample files for Chapter 8. This folder also contains detailed colour versions for the upcoming models.

Terminology

Occurrence
A copy of an existing component with the same name but with a sequenced number (e.g. Bracket:1, Bracket:2, Bracket:3, etc.).

Active component
A component that has been selected to edit (for example), and only one component in an assembly can be active.

Ground component
A component that is stationary and the first component placed into an assembly file is always grounded. Other placed components can be grounded by the user as required.

Sub-assembly
Components that are grouped together in an assembly file. An existing assembly can be placed into another assembly or a new assembly can be created from within the existing assembly file.

Constraints
Used to 'attach and fix' components to each other.

Constraint types
There are several constraint types, including:

1 **Mate** – there are three types of Mate Constraint: Plane (face), Line (edge) and Points.
   (a) Planes on one part can be constrained to planes on another part.
   (b) Lines on one part can be constrained to lines on another part.
   (c) Points on one part can be constrained to points on another part.

2 **Angle** – used to define and angle between planes of two individual parts.

3 **Tangent** – used to constrain planes, cylinders, spheres and cones. At least one of the selected faces should be curved.

4 **Insert** – can only be used on components with circular edges. This constraint aligns the centre lines of each part and applies a mate constraint to the planes defined by the circular edges.

5 **Motion** – allows animation of gears, pulleys, rack and pinions, etc. There are two types of motion constraint: rotation and rotation/translation.

6 **Translational** – specifies the intended relationship between faces.

Exercise H1: angle bracket

Use the dimensioned sketch below to create both parts, then create the angle bracket assembly and save when the assembly is complete.

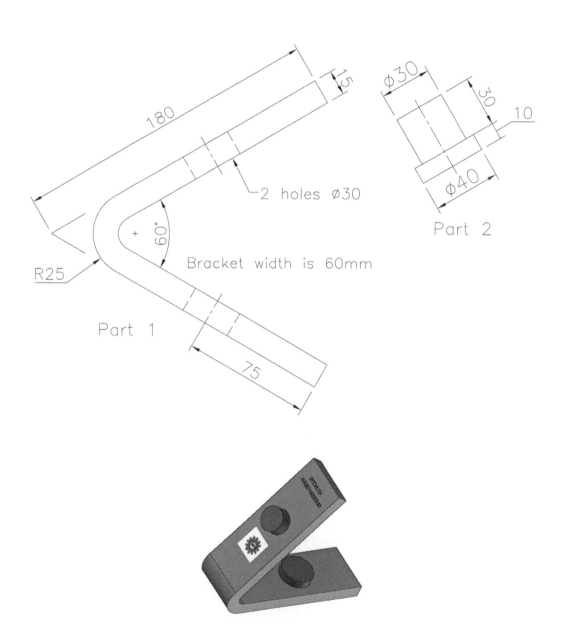

Exercise H2: fork connector and rod

1 Another two-part assembly.

2 Use the dimensioned sketch to create the parts, then create a fully constrained assembly with the bar 'central' to the connecting rod.

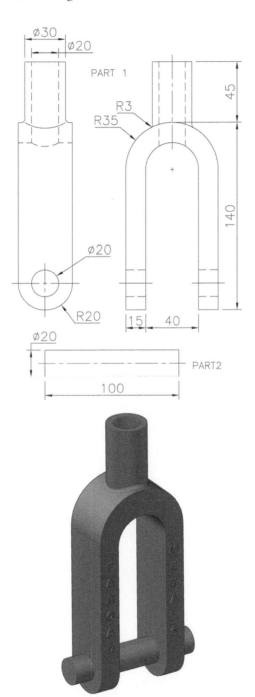

Exercise H3: back panel

Create an assembly of the back panel that consists of the following dimensioned parts, and save when complete:

1   Part 1: Back Panel (1 of).

2   Part 2: Retained Rod (1 of).

3   Part 3: Washer – 4mm deep (4 of).

4   Part 4: Holding Screw (simulated) (4 of).

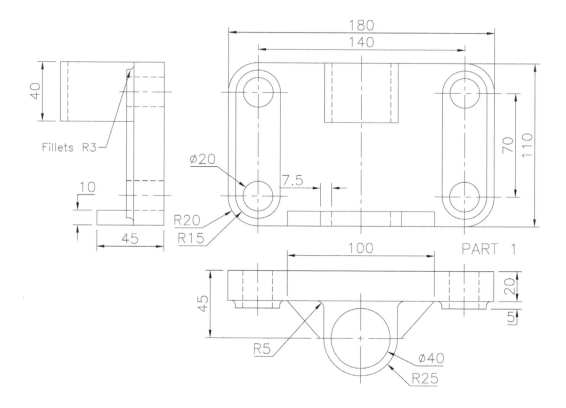

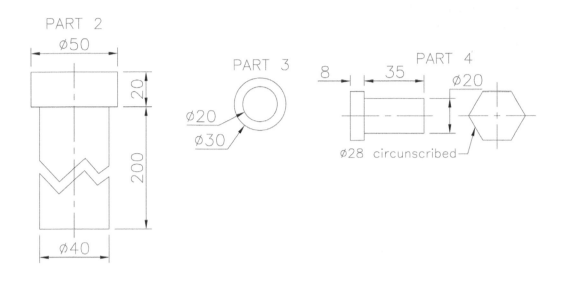

PART 2

Ø50

20

200

Ø40

PART 3

Ø20

Ø30

PART 4

8

35

Ø20

Ø28 circunscribed

Exercise H4: pointer

The four parts required for this assembly are:

1   Part 1: Holding Rod (1 of).
2   Part 2: Guide Plate (2 of).
3   Part 3: Pointer (1 of).
4   Part 4: Operator Lever (1 of).

Using the displayed dimensioned sketches, create each part (colour as required) and save, then complete the assembly.

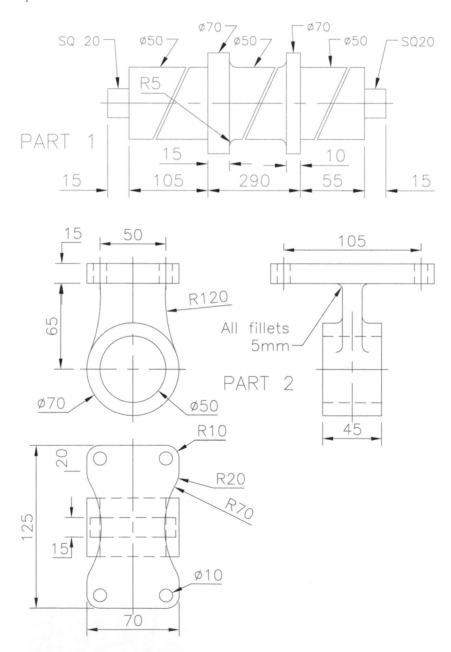

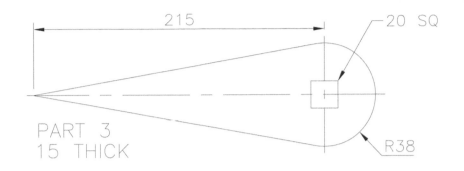

215

20 SQ

PART 3
15 THICK

R38

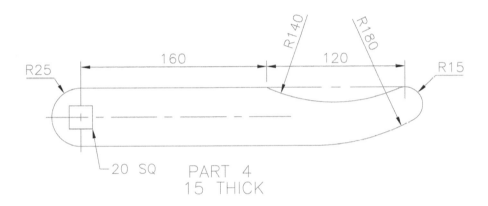

R140

R180

R25

160

120

R15

20 SQ

PART 4
15 THICK

Exercise H5: wheel and handle

There are three parts in this assembly:

1   Part 1: Wheel (1 of).

2   Part 2: Handle (1 of).

3   Part 3: Retaining Screw (1 of).

The sketch below displays the assembly:

1   as a front view; and

2   as a sectional end view through the vertical centre line.

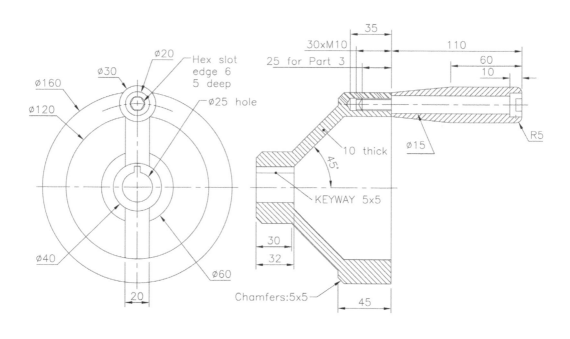

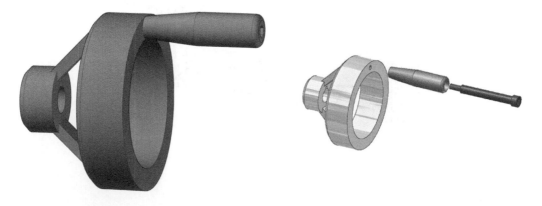

Exercise H6: point marking device

There are four parts (all 1 of) to this 'partial' assembly, these being:

1   Part 1: Base.

2   Part 2: Lever.

3   Part 3: Pin.

4   Part 4: Pivot.

Create each part and save, then complete an assembly of the device.

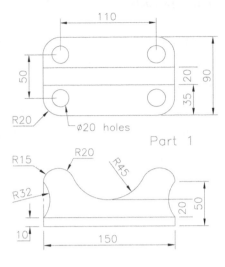

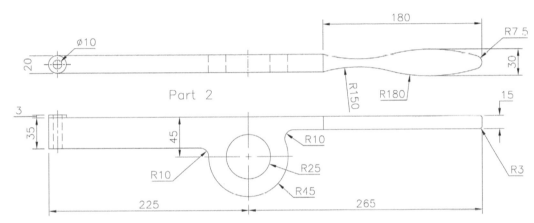

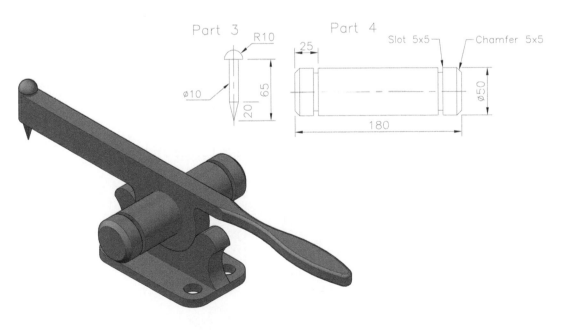

Exercise H7: steam venting valve

There are three separate parts to this assembly:

1 Part 1: Valve Body (1 of).

2 Part 2: Flange (2 of, upper and lower).

3 Part 3: Bolts (8 of, M10 square headed).

Note: I have not included any washers in this assembly, but you can include these if you want.

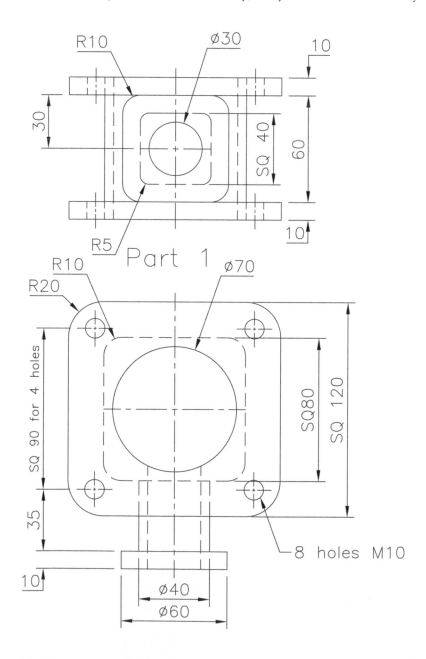

40 octagon edge

20

10

Part 2

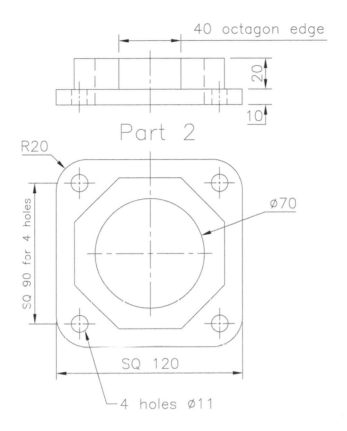

R20

SQ 90 for 4 holes

Ø70

SQ 120

4 holes Ø11

Part 3

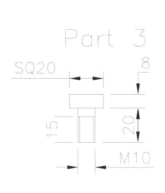

SQ20

8

15

20

M10

Exercise H8: process adaptor

The process adaptor has the following five parts to be assembled:

1   Part 1: Back Plate (1 of).

2   Part 2: Body (1 of).

3   Part 3: Cap (1 of).

4   Part 4: Cap Top (1 of).

5   Part 5: M10 BOLT (4 of, Library with M10 ISO bolt).

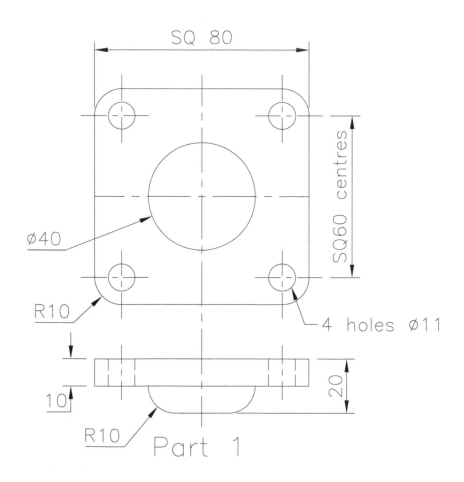

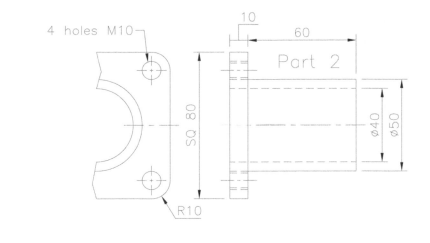

4 holes M10

10 60

Part 2

SQ 80

R10

Ø40 Ø50

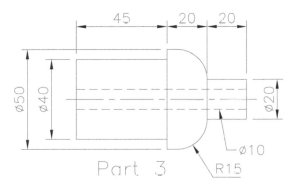

45 20 20

Ø50 Ø40

Ø20

Ø10

Part 3 R15

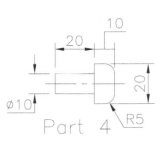

10 20

20

Ø10

Part 4 R5

Exercise H9: roller spindle

The individual parts for the roller spindle assembly are:

1   Part 1: Frame (1 of).

2   Part 2: Spindle (1 of).

3   Part 3: Rod (1 of).

4   Part 4: M12 Washer (1 of from Library).

5   Part 5: M12 Nut (1 of from Library).

6   Part 6: Plug (1 of).

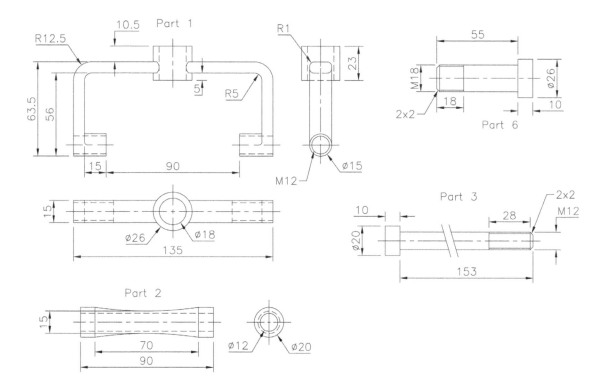

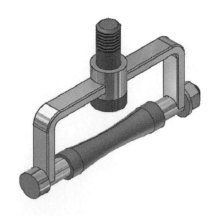

Exercise H10: swinging arm rig

The parts for this assembly are:

1   Part 1: Base (1 of).

2   Part 2: Upright (1 of).

3   Part 3: Arm (1 of at 15 degrees to the horizontal).

4   Part 4: Arm Pin (1 of ∅10, length to suit).

5   Part 5: M14 Bolt (4 of, length 50mm).

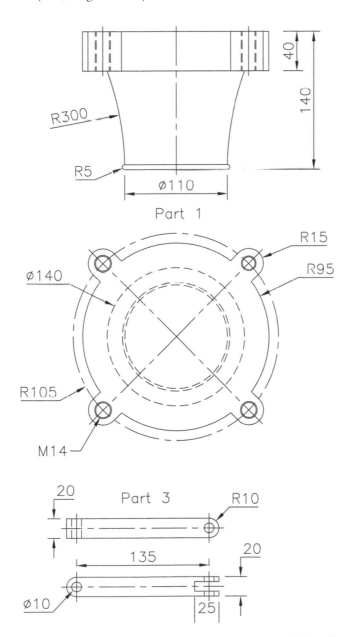

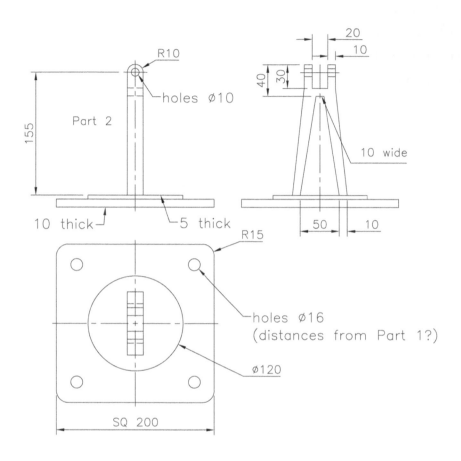

R10

holes ⌀10

Part 2

155

10 thick

5 thick

20

10

40

30

10 wide

50

10

R15

holes ⌀16
(distances from Part 1?)

⌀120

SQ 200

Exercise H11: 90-degree rocker arm

The parts for this assembly (all 1 of) are:

1   Part 1: Base.

2   Part 2: Fulcrum Pin.

3   Part 3: Bush.

4   Part 4: Arm in vertical position.

5   Others: M22 Hex Nut and Washer.

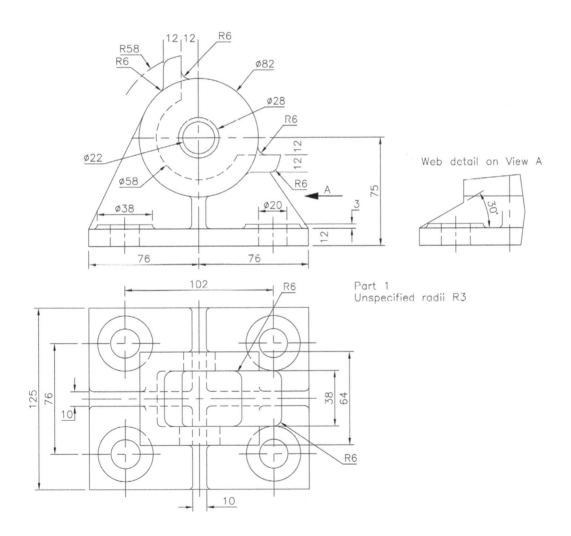

Web detail on View A

Part 1
Unspecified radii R3

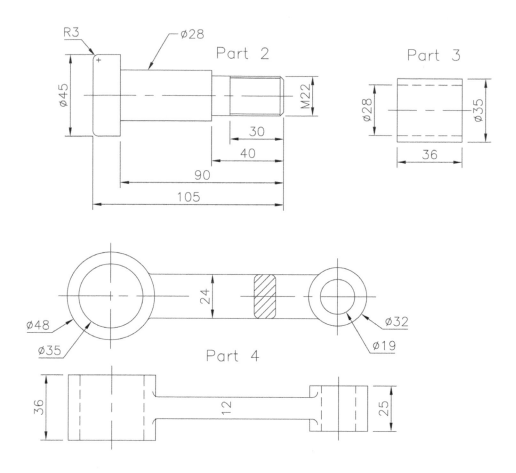

Part 2

Part 3

Part 4

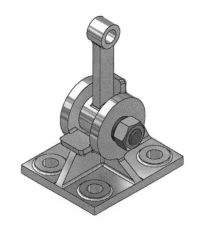

Exercise H12: shackle

The parts for the shackle assembly (material as appropriate) are:

1   Part 1: Body (2 of).

2   Part 2: Wheel (1 of).

3   Part 3: Axle Bolt (1 of and M12 nut).

4   Part 4: Securing Bolt (2 of and 2 × M8 nut).

5   Part 5: Shackle Ring (1 of).

6   Part 6: Washers to suit your decision.

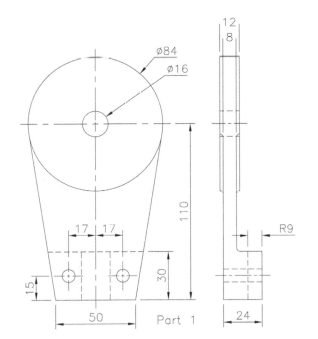

Part 1

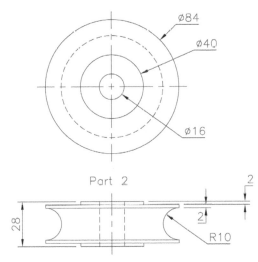

Part 2

Unspecified radii R2

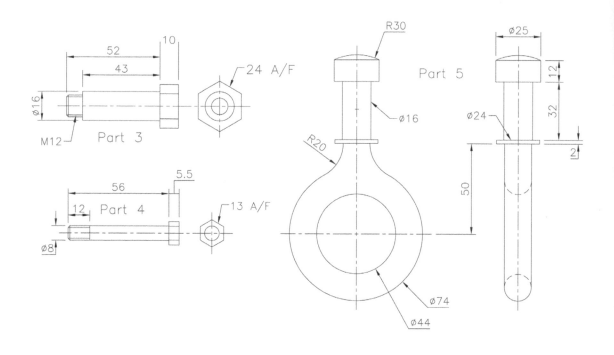

52
43
10
Ø16
M12
Part 3
24 A/F

56
5.5
12   Part 4
Ø8
13 A/F

R30
Part 5
Ø16
R20
Ø74
Ø44

Ø25
12
Ø24
32
50
2

1 Drawing layouts

1   The object of the exercises in this chapter is to create drawing layouts using already created part models and assemblies.

2   There is no new model creation work required.

3   Users should either:
   (a) use their own standard sheet layout; or
   (b) use the Inventor® standard sheet.

4   The procedure for each exercise is:
   (a) Start each exercise by opening an *already created and saved* part model or assembly file.
   (b) Open *a new metric .idw file*, this being:
      • the Inventor standard .idw file;
      • the Inventor ISO.iwd or GB.idw file; and
      • your own standard .idw file.
   (c) Use the information given in each exercise to produce the drawing layout and add a Parts List for assembly layouts.
   (d) Add any refinements of your own to enhance the layout.
   (e) Save the completed layout file.
   (f) Use your discretion as required.

5   *Note*:
   (a) It is user preference as to whether to display the drawing layout in First or Third Angle, this being dependent on individual/company policy and the user standard sheet being used.
   (b) The exercises will have both First and Third Angle layouts, but the user does not need to adhere to the exercise requirements.

Terminology

Base view

1   This is the first view created from the part or assembly model.

2   The user can select whether to display a Top, Front, Right view.

Projected view
An orthogonal or isometric view projected from the base view.

Auxiliary
An additional view projected from an edge or line in the parent view.

Section
Full, half, offset and aligned section views can be created from a parent view and are automatically aligned to the parent view.

Detail
Detailed views of specific 'parts' of a parent view can be created and are not aligned to the parent view.

Broken view
Used when the component view exceeds the length of the drawing.

Break out view
Removes a user-defined area to expose 'hidden' parts or features in the existing drawing.

Exercise I1: drain plate cover (E1)

1   Open the drain plate cover part model from Exercise E1.

2   Open a new standard .idw file.

3   Create a layout in First Angle projection with the following views:
    (a) top, front and end view;
    (b) section end view (selecting your own section line);
    (c) an auxiliary elevation;
    (d) any detail view adding dimensions; and
    (e) a 3D view of the model.

4   Save the layout if required.

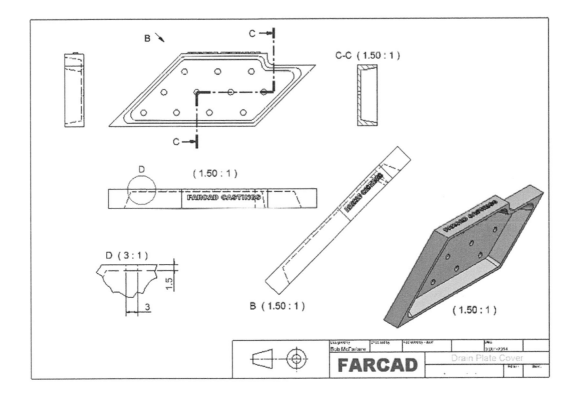

Exercise I2: connector (E2)

1   Open the connector part model from Exercise E2.

2   Open a new standard .idw file.

3   Create a layout in Third Angle projection with the following views:
   (a)   top, front and end view;
   (b)   section end view (selecting your own section line);
   (c)   detail views as appropriate; and
   (d)   a 3D view of the model.

4   Save the layout if required.

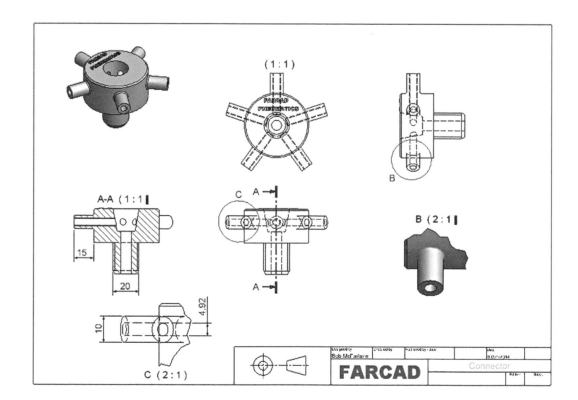

Exercise I3: ducting corner (F3)

1 Open the part model of the ducting corner from Exercise F3.

2 Open a new standard .idw file.

3 Create a First Angle layout with the following views:
   (a) top, front and end views;
   (b) a sectional end view (selecting your own section line);
   (c) an auxiliary plan view;
   (d) a 3D view of the model; and
   (e) a detail view with dimensions.

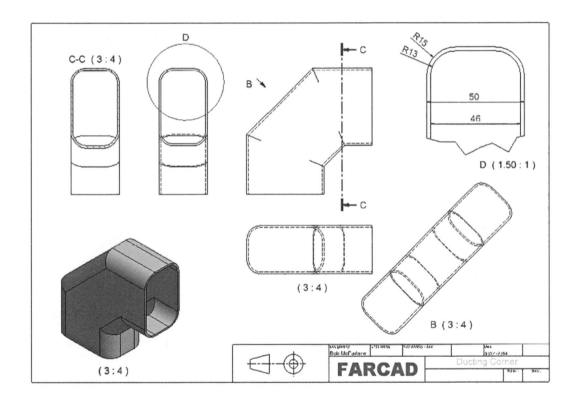

Exercise I4: pinion gear (F13)

A First Angle projection layout to be created with:

1   top and front views;

2   a sectional elevation through an arbitrary section line CC;

3   two detail views A and B with dimensions added;

4   a 3D view of the component; and

5   a detail view D in 3D mode.

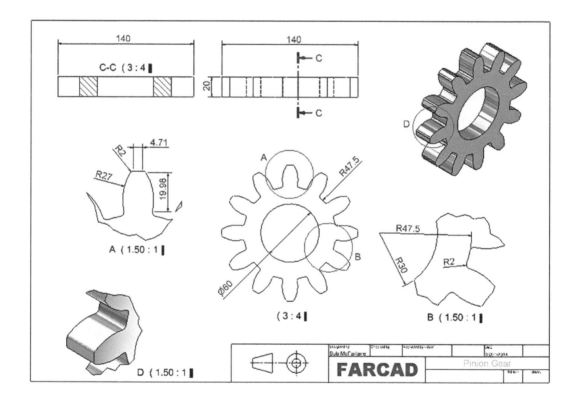

Exercise I5: spout (F20)

Create a First Angle projection layout with the following displayed:

1 plan, elevation, end elevation and a 3D view;

2 a sectional end elevation through model centre line;

3 detail B at 1.5:1 with dimensions;

4 detail C at 2:1 with hole data added; and

5 several dimensions added to the three main views.

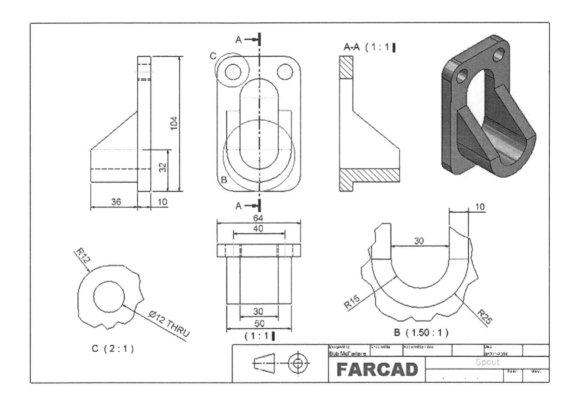

Exercise I6: fan motor housing (F21)

A Third Angle projection layout to be created with:

1   a plan, elevation, end elevation and 3D view;

2   a sectional end elevation through model centre line;

3   several detail views with some dimensions; and

4   an item of text.

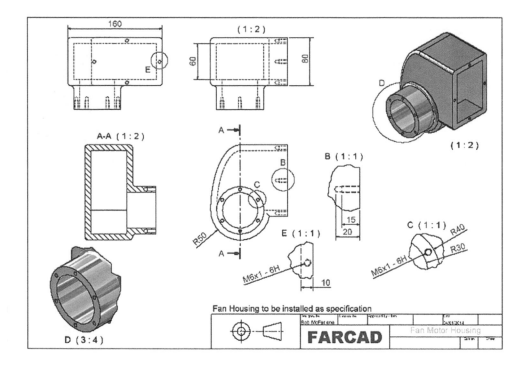

Exercise I7: Swiss mechanism drive (G10)

Create a layout to your own specification.

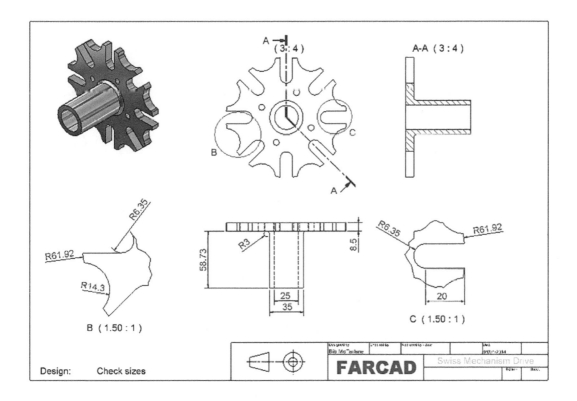

Exercise I8: brake shoe (G20)

Create a layout of the brake shoe model to your own specification.

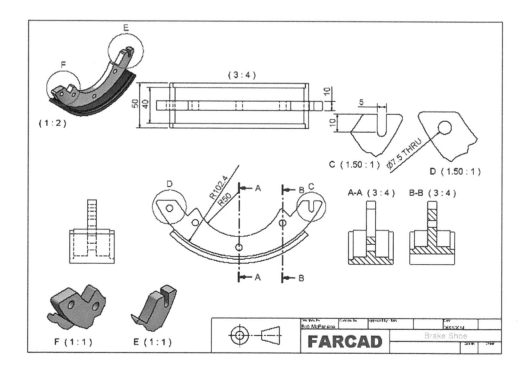

Exercise I9: back panel (H3)

This is an assembly layout:

1   Open the saved back panel assembly from Exercise H3.

2   Using a suitable drawing sheet, create in First Angle projection a layout (at a scale of your choice) to include:
    (a) three traditional orthographic views;
    (b) a 3D view of the assembly;
    (c) any suitable section view;
    (d) any suitable detail view with some dimensions;
    (e) balloons positioned to suit; and
    (f) a parts list, 'adjusted' to your requirements.

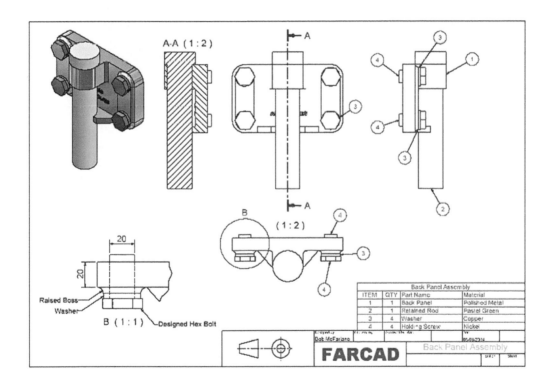

Exercise I10: steam venting valve (H7)

A traditional type layout to be created to your specification.

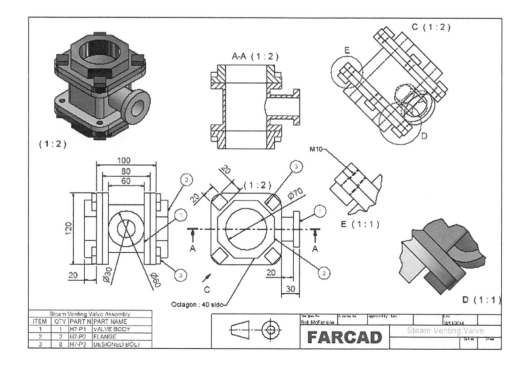

Final note

1   The creation of a drawing layout should now be familiar (and perhaps tedious), and the 10 exercises should have familiarised the user with the process.

2   The user can create their own drawing layouts using the models and assemblies that have previously been created.

3   Drawing layouts will also be required in several of the chapter exercises still to be completed.

J   Sheet metal design

1   There are many processes in sheet metal design (fabrication), including:
   (a)   stamping;
   (b)   drawing;
   (c)   punching;
   (d)   cutting;
   (e)   rolling; and
   (f)   other complex operations.

2   Sheet metal work can be considered as:
   (a)   a metal blank folded into a finished shape; and
   (b)   'thin plate' work (i.e. less than 1 inch thick).

3   With Inventor®, the general procedure with sheet metal work is to:
   (a)   create a sheet metal part using faces;
   (b)   create key features of the sheet metal part;
   (c)   add/create corner seams;
   (d)   cut shapes from the faces;
   (e)   add standard features (e.g. chamfers and fillets);
   (f)   create a flat pattern model of the part; and
   (g)   create a drawing layout of the folded part and flat pattern.

4   The procedure for the sheet metal exercises is:
   (a)   Open *a new metric sheet metal (mm) .ipt* file.
   (b)   Using Applications, check that the sheet metal environment is active.
   (c)   Set Styles (and Save) as required for the Sheet, Bends, Rounds and Corners.
   (d)   Use the information given in each exercise to create the part model and/or layout as required.
   (e)   Save the completed models/layouts.

Terminology

Thickness
This is the thickness of the flat stock used to create the sheet metal part.

Unfold method
This is the method used to calculate bend allowance – material stretching during bending.

Transition

1   This is the control of the intersection of edges across a bend in the flattened sheet.

2   The transitions available with Inventor include:
(a) none;
(b) intersection;
(c) straight line; and
(d) arc.

Relief

1   This is a small notch cut next to the end of a bend if the bend does not extend the full width of an edge.

2   The notch can have a straight or round end.

Corner

1   Occurs where three faces meet and corner relief is added to the flat sheet before bending.

2   There are different types of corner relief, including:
(a) round;
(b) square;
(c) tear; and
(d) trim to bend.

Flange

A sheet metal flange is a simple rectangular face created from an existing face.

Hem

1   Used to eliminate sharp edges or strengthen an open edge of a face.

2   Hem types include:
(a) single;
(b) double;
(c) teardrop; and
(d) rolled.

Bend

Connects two faces with a straight, angled or curved face.

Fold

Allows flat faces to be folded along sketched lines on a face.

Cut

1   Used to 'extrude' a cut shape usually created on a sketched face.

2   The cut can 'cross' bends and folds.

Punch

Used to create cut-out shapes in flat faces but requires specifically designed iFeatures on sketches.

Exercise J1: inspection cover

1   With *a new metric sheet metal (mm) .ipt file*, refer to the orthographic layout below and create an open-topped inspection cover (made for 1.6mm aluminium) using the 'overall' dimension given (remember that this diagram does not take any account of bends, tabs, etc.).

2   Each side is to have a 30mm wide mounting flange at 90 degrees to the side.

3   The following features are also required:
(a)   10 Ø10mm holes positioned to your specification on the flanges;
(b)   a single Ø40mm inspection hole in the 'centre' of the top face;
(c)   the long flanges to be chamfered 10 × 10 at the corners;
(d)   the shorter flanges to have an R10 corner fillet; and
(e)   four corner seams to be added.

4   Save the model when complete, still with the gaps between sides – we may refer to it in another section.

5   When the model of the inspection cover model is complete:
(a)   Create a flat pattern of the enclosure and save.
(b)   Create a drawing layout (either in First or Third Angle projection) to display:
   •   three 'basic' orthographic views;
   •   a 3D view of the enclosure; and
   •   the flat pattern with dimensions.

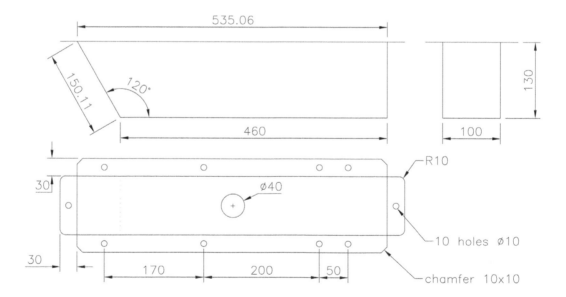

Note: Dimensions do not take account of bends/folds, etc.

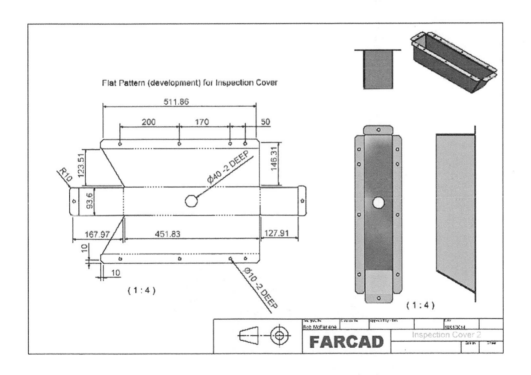

Flat Pattern (development) for Inspection Cover

511.86

200   170   50

123.51   R10   146.31

Ø40 -2 DEEP

93.6

167.97   451.83   127.91

10

10

Ø10 -2 DEEP

(1 : 4)

(1 : 4)

Rob McFarlane

FARCAD

Inspection Cover 2

Exercise J2: sheet metal FOLLY

1   A sheet metal 'FOLLY' model has to be created from 1.6mm brass using the basic drawing information given below.

2   The 'free ends' of the four sides of the folly have to be 'hemmed' as follows:
   (a) single, length 20, gap 2;
   (b) teardrop, radius 2, angle 190;
   (c) rolled, radius 3, angle 290; and
   (d) double, length 20, gap 2.

3   Create and save the FOLLY sheet metal part model.

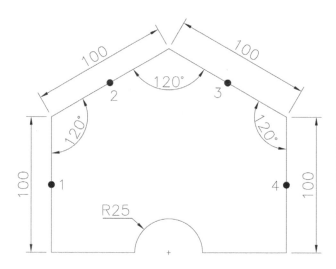

Side 1   Height 50, Single hem
Side 2   Height 60, Teardrop hem
Side 3   Height 70, Rolled hem
Side 4   Height 80, Double hem

All sides are 'inclined' at
95° to the horizontal

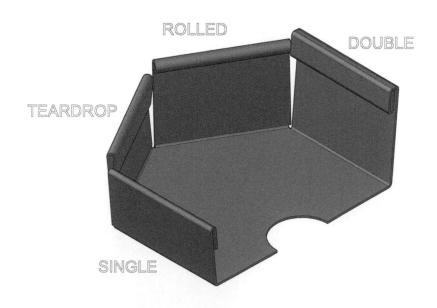

Exercise J3: cable grip

1   A nylon cable grip has to be created from the following information:
    (a)   Two vertical 'plates' with sizes $200 \times 80 \times 1.6$ and $150 \times 80 \times 1.6$.
    (b)   The plates are 50mm apart at the 80mm edge.
    (c)   The longer plate has a 110×20 cut slot.
    (d)   The shorter plate has four holes cut in it.
    (e)   Each plate has a fold along a given line for 30 degrees.

2   Using the information given above and the drawing dimensions below:
    (a)   Create a sheet metal model of the cable grip.
    (b)   Create a drawing layout to include:
        •   the three traditional orthographic views at either First or Third Angle;
        •   a 3D view of the model; and
        •   any auxiliary view.

3   Using a new drawing layout, produce the flat pattern development with dimensions that will assist manufacture.

4   Remember to save all work.

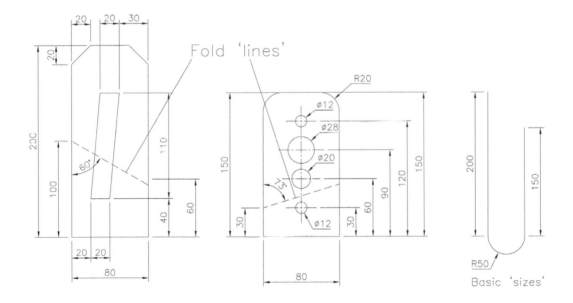

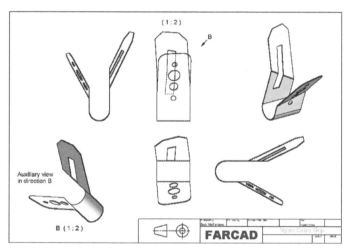

(1:2)

B

Auxiliary view
in direction B

B (1:2)

FARCAD

Nylon Cable Grip

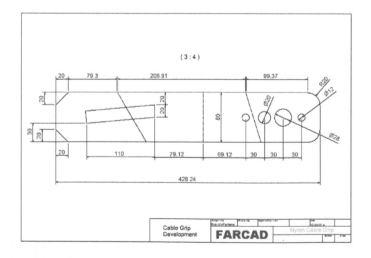

(3:4)

20   79.3   206.91   99.37   R20
Ø12
20
20   Ø20
30
80
20
Ø28
20
20   110   79.12   69.12   30   30   30
428.24

Cable Grip
Development

FARCAD

Nylon Cable Grip

Exercise J4: patterned panel

1 Information:
   (a) Aluminium plate 540×200 and 1.6mm thick.
   (b) 30mm and 50mm folds at the ends.
   (c) Various shapes cut out of the flat and ends.

2 Using the information given above and the dimensions in the sketch below, create a sheet metal model of the panel, and:
   (a) shape A pattern distances are 70 horizontal and 60 vertical; and
   (b) shape B pattern distances are 50, both horizontal and vertical.

3 Save your model when complete.

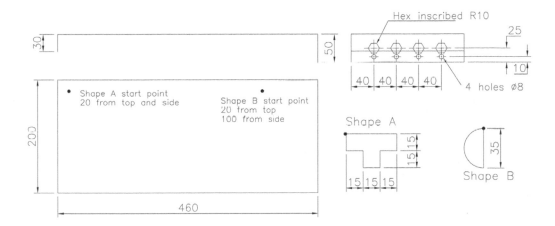

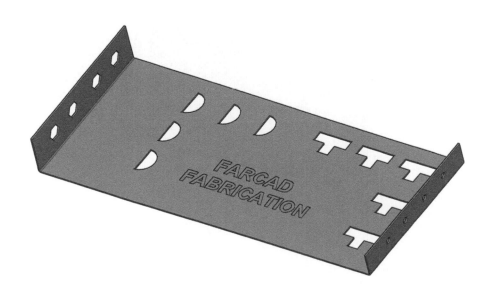

Exercise J5: herb labels

1   A company produces kitchen utensils and one of its products is a label with the name of a herb, made from different materials.

2   Information:
   (a) The plate is $100 \times 50 \times 1.6$.
   (b) The plate has:
       •   a 6×6 square cut-out at each corner; and
       •   a rolled hem (R1mm and angle 290 degrees on the four sides).
   (c) The herb name can be embossed, engraved or 'cut through'.

3   Design a herb label for the three (c) options.

EMBOSSED   ENGRAVED

CUT-THROUGH

Exercise J6: 90–degree corner V channel

1   A channel system is made from 1.6mm thick aluminium, and the X-section and plan view of part of the channel are displayed below.

2   Create a sheet metal model for this component, which is open at both ends.

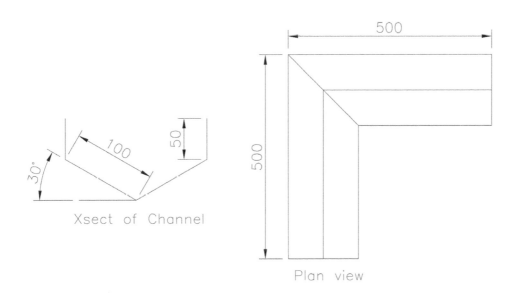

Xsect of Channel

Plan view

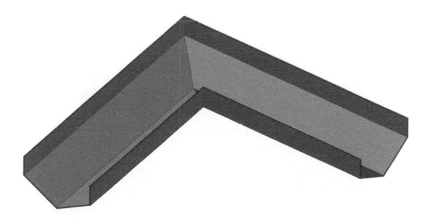

Exercise J7: bracket

Create a sheet metal bracket from 1.6mm material using the drawing information given in the First Angle orthographic views below.

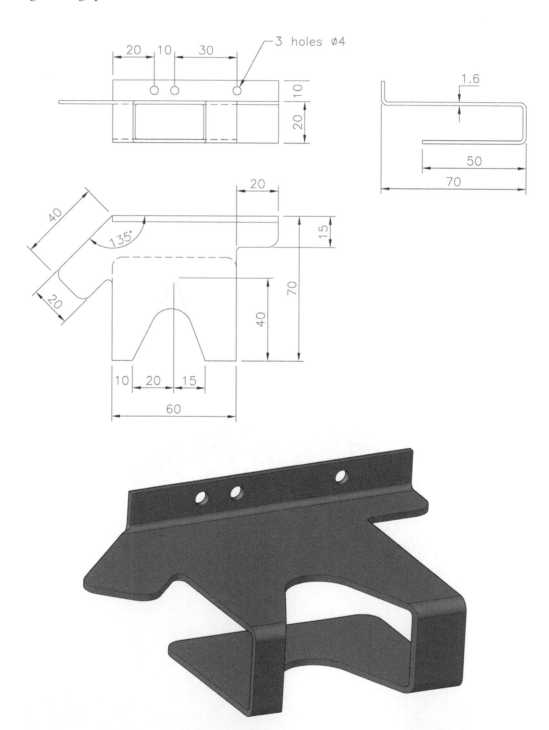

Exercise J8: vent panel

1 A computer vent panel is made from 0.5mm mild steel plate.

2 Using the orthographic drawings below, create the vent panel.

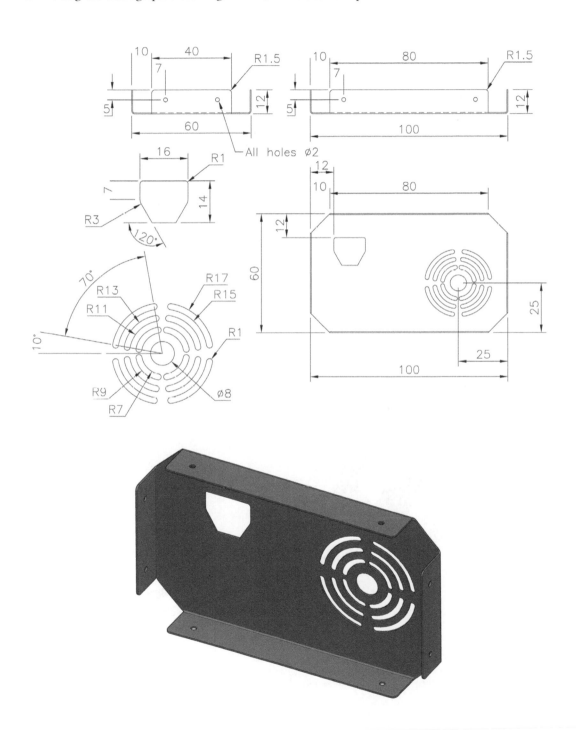

Exercise J9: brass gasket

Create a sheet metal model of the unusual shaped 0.5mm brass gasket using the dimensioned drawing below. The gasket has a 60-degree fold, as indicated.

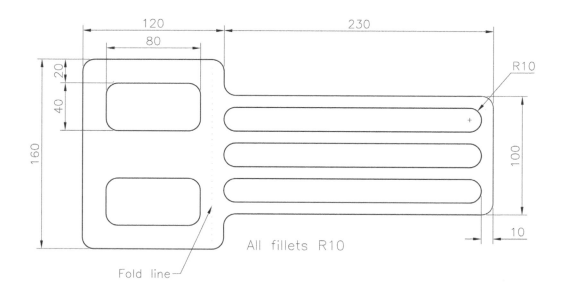

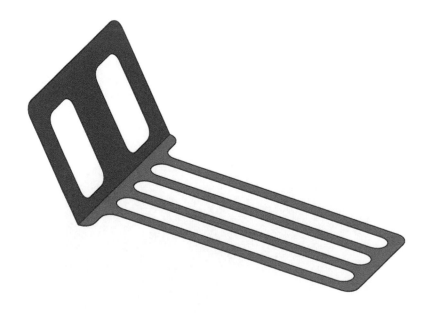

Exercise J10: printed circuit board

1   A printed circuit board has to be created from 1.6mm nylon using the drawing below for the various pathways.

2   Using the information from the drawing (and your discretion for radii, etc.), create the component, noting that all pathways, terminals, etc. are etched 0.5mm into the board.

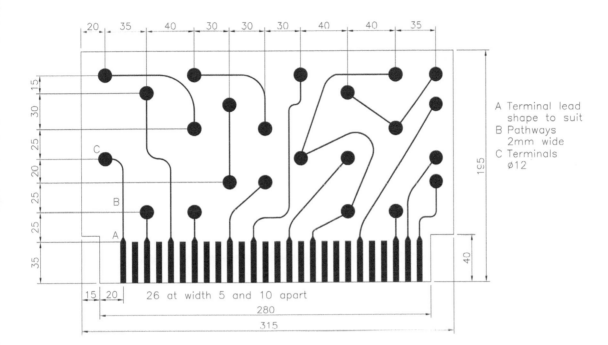

A Terminal lead
   shape to suit
B Pathways
   2mm wide
C Terminals
   Ø12

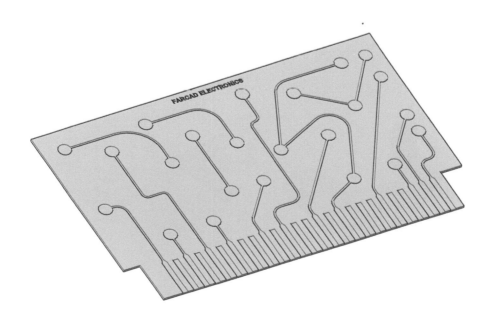

Exercise J11: connecting bracket

1 Use the two orthographic drawings to create a sheet metal model of the connecting bracket.

2 When the model has been completed, create a drawing layout in Third Angle projection with the flat pattern displayed (with dimensions) full size.

3 The material is 2.0mm thick mild steel with 2mm rounds and a 2mm inside bend radius.

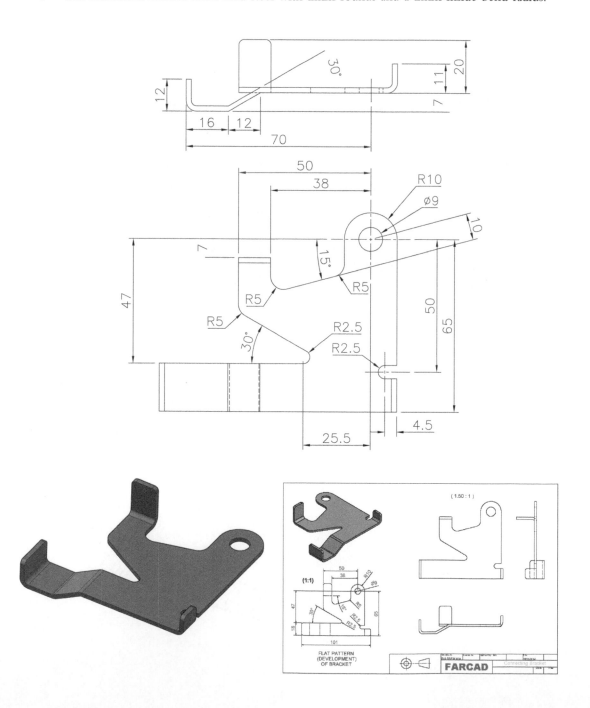

Exercise J12: square chute

1 Create the Mild Steel 0.5mm thick sheet metal chute using the drawing information given below.

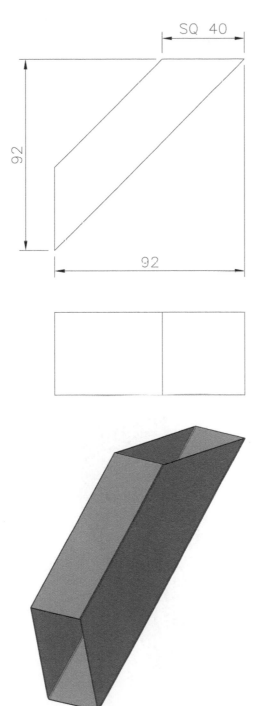

2   With a slight variation in the plan view (60-degree angle included as below), create the 'new' model.

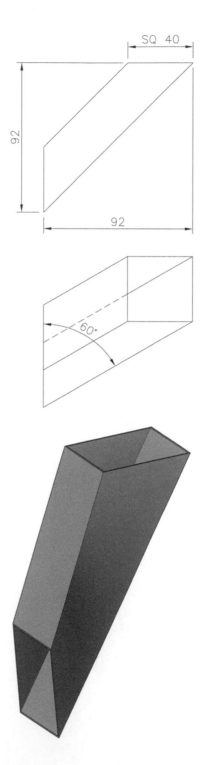

K Additionals

1 This chapter is intended to allow the user additional practice with a variety of exercises.

2 The user will be creating additional part models, assemblies and drawing layouts.

3 There is no order to the exercises (i.e. the last is not necessarily harder than the first).

4 The procedure for the exercises should (by now) be obvious to the user, so I have not included any steps − it is now your choice as to how each exercise is completed.

Exercise K1: carbide tip saw

1   Create a part model of the carbide tip saw using the dimensions given below.

2   The disc is 2mm thick and the actual tip is 5mm thick.

3   A suggested start position for the 'tooth' is indicated.

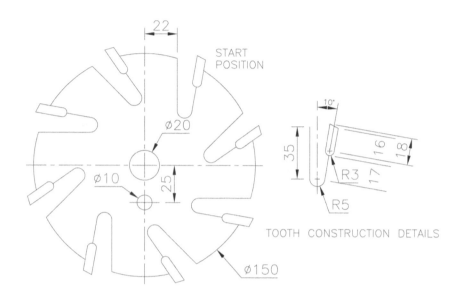

Exercise K2: model tower

1   Using the information in the sketch below, create a part model of the tower.

2   Use your discretion as appropriate and add some extra 'features'.

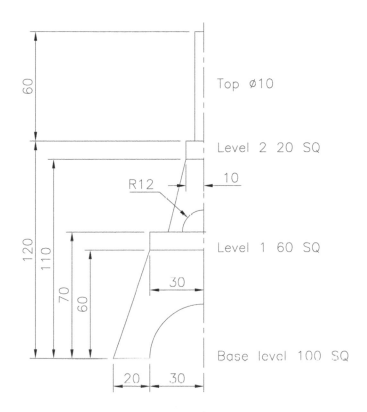

Top Ø10

Level 2 20 SQ

R12   10

Level 1 60 SQ

30

120   110   70   60   60

Base level 100 SQ

20   30

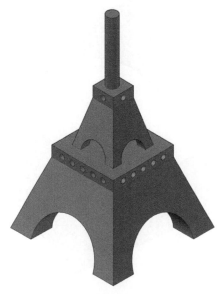

Exercise K3: flange plate

1   A $300 \times 300 \times 1.6$mm copper plate has a shape cut from it.

2   Use the diagram below to create a sheet metal part model of the plate.

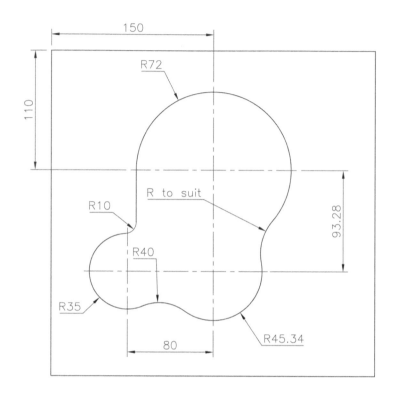

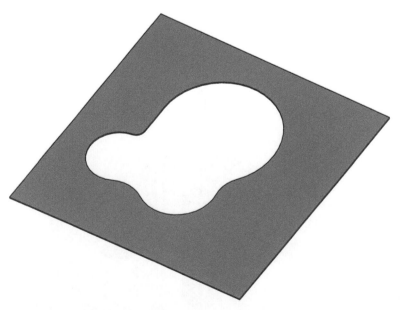

Exercise K4: the wall of steps

1 A wall has to be created from 100mm square brick with a design (your own) 'in the middle'.

2 Create a part model of the brick, then create an 8 brick × 8 brick stepped wall similar to that displayed.

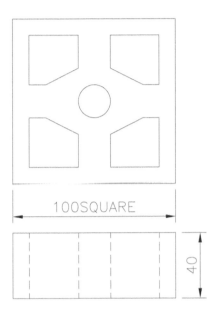

100SQUARE

40

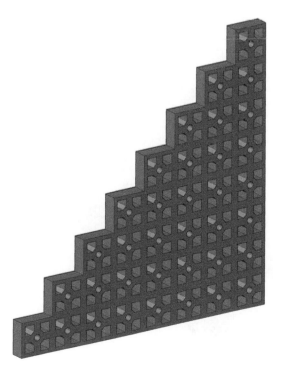

Exercise K5: pulley

1   A pulley arrangement consists of the following parts:
    (a)   Part 1: Large Pulley;
    (b)   Part 2: Small Pulley; and
    (c)   Part 3: Shaft.

2   Create each part as a part model, then create an assembly and drawing layout (in Third Angle) of the pulley arrangement.

3   Add any refinements of your choice (e.g. a parts list).

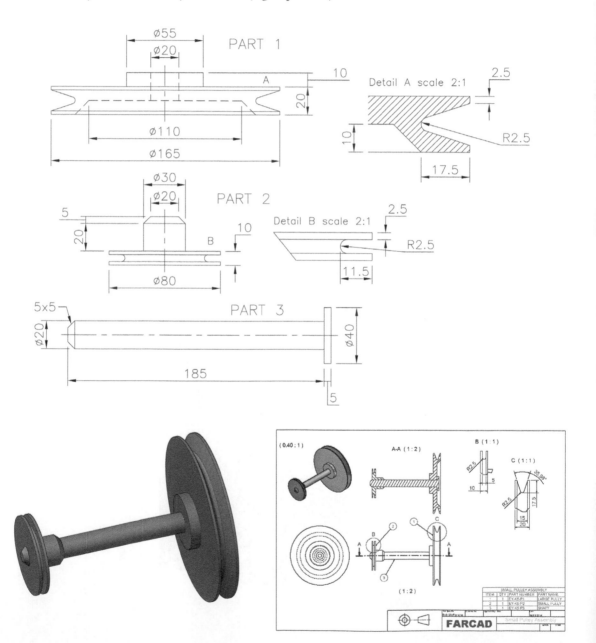

Exercise K6: the one 1

1   This exercise (problem) was posed to me by Emma Crombie, an apprentice CAD technician at Doosan Babcock in Renfrew: *Model a **single** component that will 'fit through' the three holes in a plate.*

2   Using the drawing details below, create a part model of the component and produce a typical drawing layout in First or Third Angle projection.

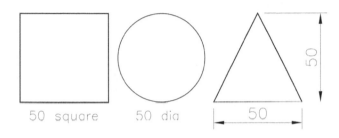

50 square   50 dia   50

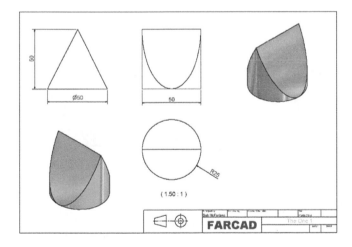

50   Ø50   50   R25

(1.50 : 1)

FARCAD   The One 1

Exercise K7: sheet metal template

1   A brass 1.6mm sheet metal plate has to be created for a special template job.

2   The template details are displayed in the drawing information below.

3   Create a sheet metal part model of the template.

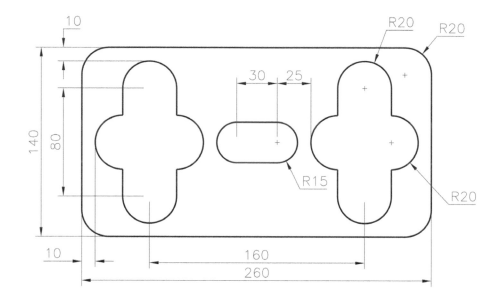

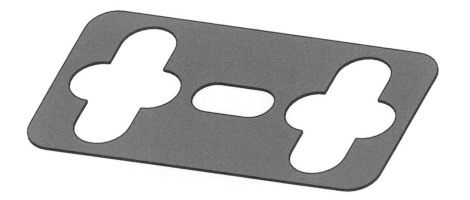

Exercise K8: connecting rod end

Using the information from the two orthographic views, create a part model and drawing layout of the connecting rod end, using discretion as appropriate.

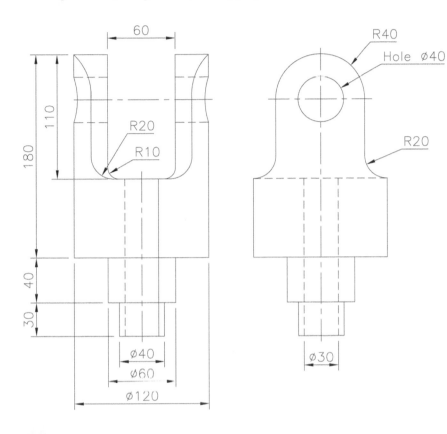

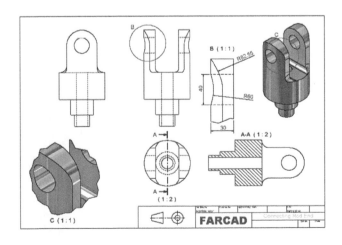

Exercise K9: garden hex 'PAGODA'

1   A garden has as a centrepiece a hexagonal-type pagoda.

2   The 'legs' have an X-section of 40×20mm and the 'path' for the pagoda legs is as shown.

3   Create a part model of the pagoda, finishing the 'top' to your own design.

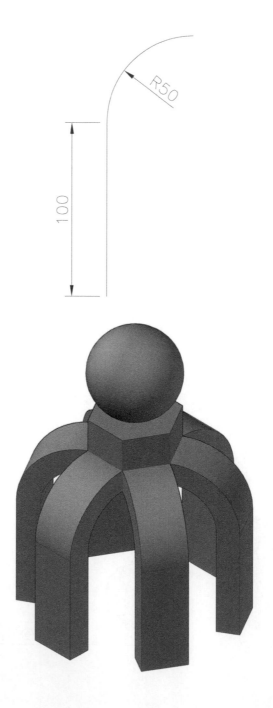

Exercise K10: pipe connector

Use the drawing information to create a part model of the component.

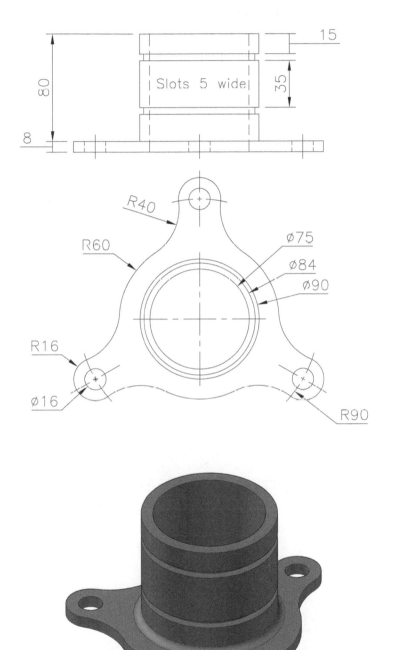

Exercise K11: tool holder

1   The dimensions for the three parts of a simple tool holder are displayed below.

2   Create a drawing layout of the assembly, displaying views of your own choice.

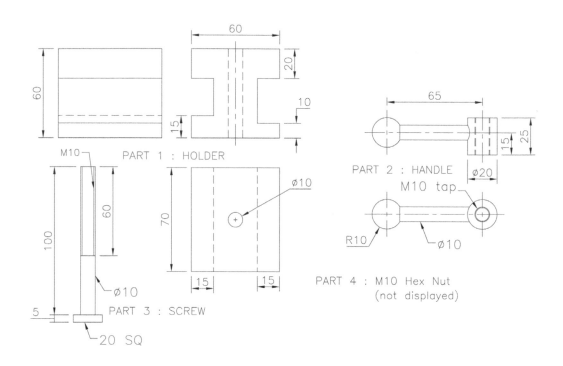

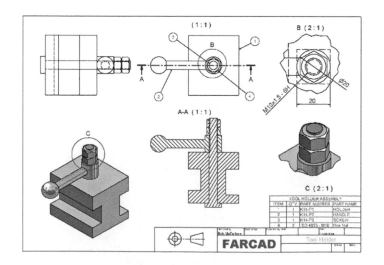

Exercise K12: pipe flange

Create a part model and drawing layout in First or Third Angle projection for the pipe flange using the drawing details given.

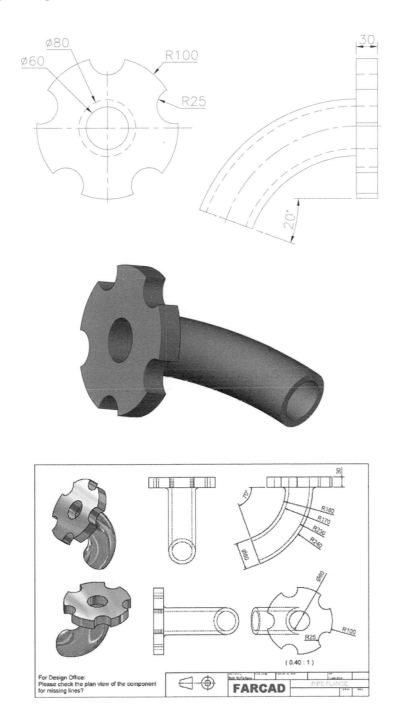

Exercise K13: blue chrome computer link

1   A special computer link is made from 4mm thick blue chrome material.

2   The link can be considered as being created from five parts, each part being 'built' on top of the edges of the previous part.

3   The diagrams below give the basic dimensions for each part.

4   Create a part model (not an assembly) of the component, and then create a detailed drawing layout of your choice.

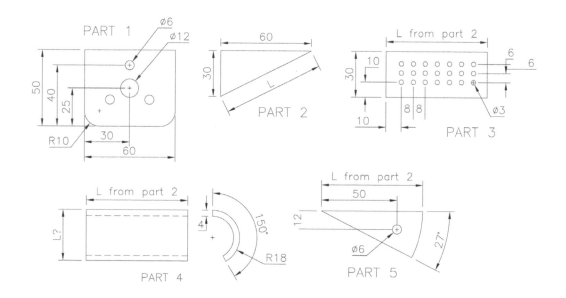

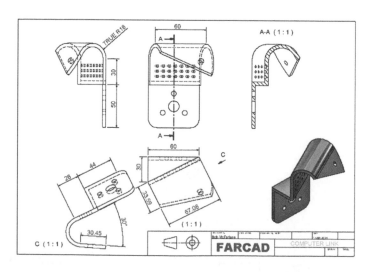

Exercise K14: jet engine exhaust

1   The special project department of a well-known aero-engine company has been tasked with producing a model exhaust system for a new engine project.

2   The model produced has an exhaust 'pipe' of 80mm diameter, which reduces to 50mm diameter over a 50mm length, and the material used has a thickness of 5mm.

3   The final exhaust is either:
   (a)   a single 'pipe' turning through an angle of 80 degrees to the horizontal;
   (b)   a twin exhaust turning through the 80 degrees; or
   (c)   a triple cascade exhaust system with the same angle of turn.

4   Create part models of each exhaust system that could aid the CFD department.

Exercise K15: angle plate

Using the information below (and your discretion), create a part model of the angle plate.

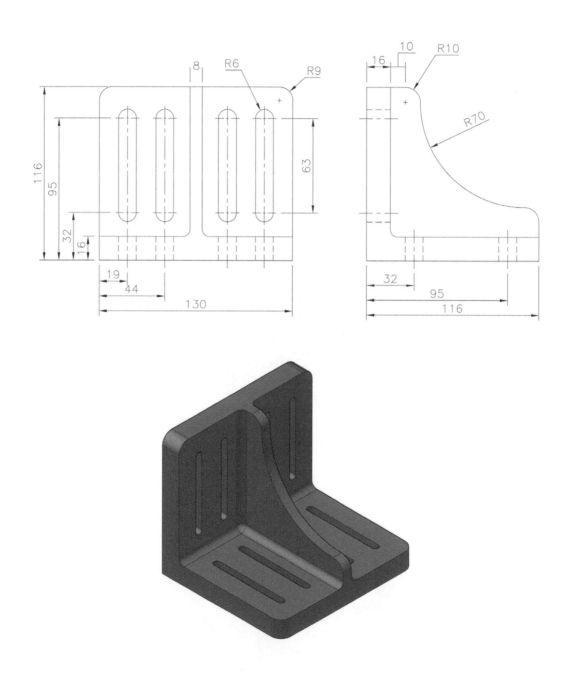

Exercise K16: V support

Using the drawing information displayed, create a part model of the V Support.

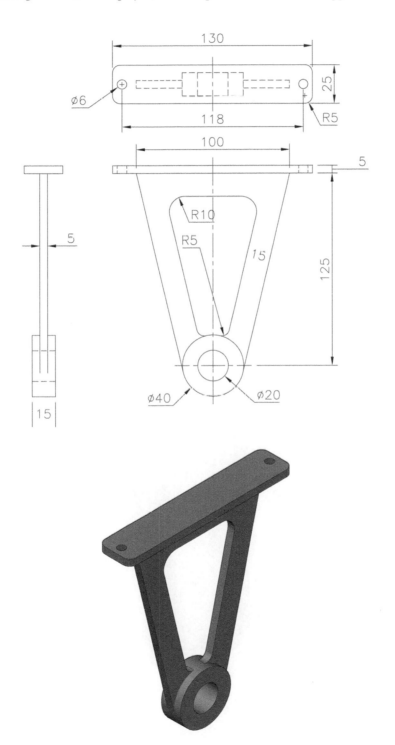

Exercise K17: setting block

Use the two drawing views to create the part model.

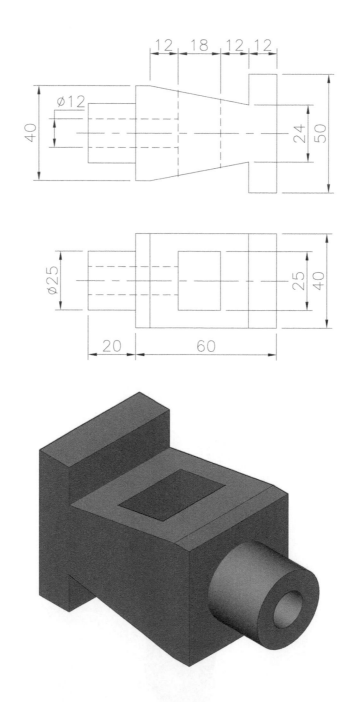

Exercise K18: wall bracket

Create a part model from the drawings given.

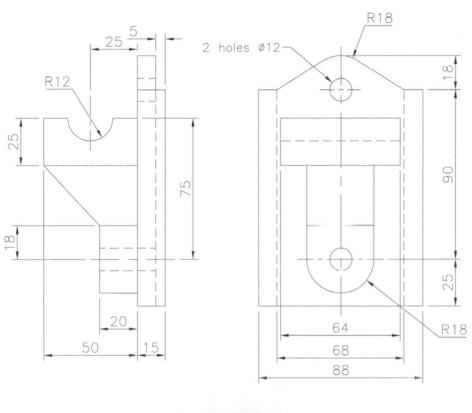

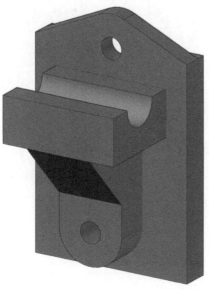

Exercise K19: support

Create a part model using the drawing information given.

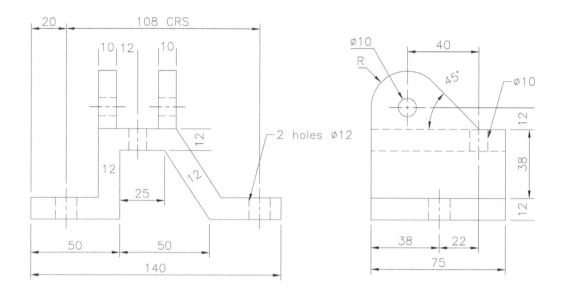

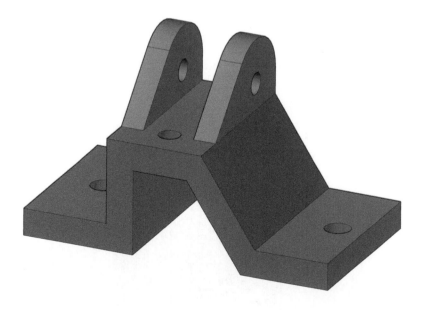

Exercise K20: customised I beam

Create a part model of the I beam using the drawing details displayed.

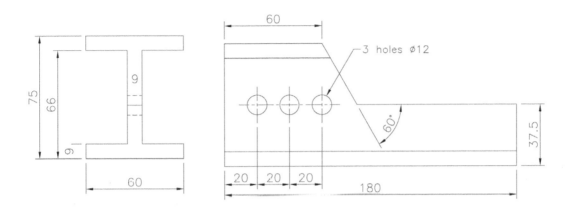

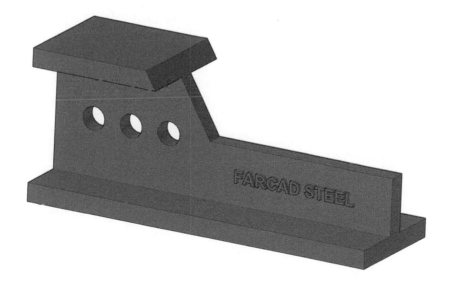

L   Presentations

1   Once an assembly has been created, it is possible to:
   (a) create an exploded view of the assembly;
   (b) create an animation of the assembly with linear and/or rotational motion; and
   (c) save the animation for future recall.

2   To create the above options, a presentation file (extension *.ipn*) is used.

3   As presentation files are used with assemblies, the Chapter 8 H exercises will be used (i.e. there are no new assemblies to create in this chapter).

4   In each exercise, the process is the same:
   (a) Start each exercise with *a new metric standard (mm) .ipn* file.
   (b) Create a view (open file) of the assembly to be animated.
   (c) Decide on automatic or manual explosion, and:
      •   if automatic, set the distance with create trails on or off; and
      •   if manual:
         –   determine the x, y, z axes orientation; and
         –   apply appropriate tweaks to the assembly components.
   (d) Review the exploded view and animation.
   (e) Save the animation, if required.

5   A number of basic animations that have been created can be accessed on the companion website (www.routledge.com/cw/mcfarlane), in the folder of sample files for animations.

6   Take a look at some completed .ipn files on the companion website (www.routledge.com/cw/mcfarlane), in the folder of sample files for Chapter 12. Colour versions for the models have also been uploaded.

Terminology

Direction

1   This is the orientation of the x, y, z axes relative to a component in the assembly.

2   This will then determine the x, y and z tweaks.

Tweak

Moves and/or rotates parts of the assembly relative to the axes orientation.

Explosion

1   This separates the parts in an assembly by a user-determined distance using tweaks.

2   The options are:
(a) manual – the user selecting the parts and entering the distance to be moved; and
(b) automatic – the assembly is automatically exploded by a given entered value.

Animation

Creates an animation of the assembly and can be saved as a *.AVI* file.

Trail

Visible lines that show how the parts of the assembly are being exploded.

Motion

Used to play the animation with several options:

1   forward by tweak – drives the animation forward, one tweak at a time;

2   forward by interval – drives the animation forward one interval at a time;

3   reverse by tweak;

4   reverse by interval; and

5   play forward – plays the animation forward.

Exercise L1: back panel (H3)

With the standard metric (mm) .ipn file, create a view of the back panel assembly with:

1   automatic explosion;

2   distance set to 50mm with create trails on;

3   play the animation; and

4   if the animation is suitable, save (with suitable names):
    (a)   the presentation file (e.g. PresentationL1-H3); or
    (b)   an .AVI file (e.g. AnimationL1-H3).

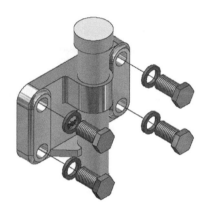

Exercise L2: pointer (H4)

With the standard metric (mm) .ipn file, create a view of the pointer assembly with:

1   automatic explosion;

2   distance set to 200mm with create trails on;

3   play the animation; and

4   if the animation is suitable, save the presentation and .AVI files with suitable names.

Exercise L3: wheel and handle (H5)

With the standard metric (mm) .ipn file, create a view of the wheel and handle assembly with:

1   manual explosion;

2   both rotational and linear motion included with create trails on;

3   play the animation; and

4   if the animation is suitable, save both presentation and .AVI files.

Note: For the rest of the exercises in this chapter, I will state what the explosion method was and what motion was applied to the animation prior to saving the presentation and .AVI files.

Exercise L4: point marking device (H6)

1   Manual explosion with trails on.

2   Rotational for the lever and pin 'about' the pivot.

3   Linear for the lever, pin and pivot from the base.

Exercise L5: steam venting valve (H7)

1   Automatic explosion with 50 distance and trails on.

2   Clockwise rotation of 20 degrees.

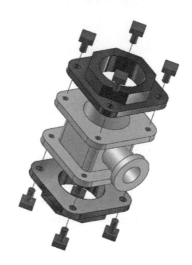

Exercise L6: process adaptor (H8)

1   Automatic explosion with 50 distance and no trails.

2   Display as wire frame with X-ray ground shadow (for a change), but animation saved with shaded display.

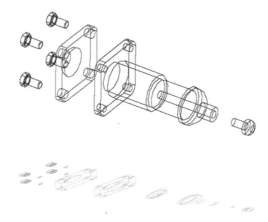

Exercise L7: roller spindle (H9)

1   Manual explosion with trails.

2   All parts moved away from the FRAME by varying distances.

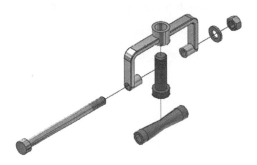

Exercise L8: swinging arm rig (H10)

1   Manual explosion with trails.

2   The swinging arm rotated about the ARM PIN.

3   The arm pin moved away from the ARM.

4   The swinging arm moved away from the UPRIGHT.

5   All parts moved away from the BASE by varying amounts.

Exercise L9: 90-degree rocker arm (H11)

1   Manual explosion with trails.

2   Arm rotated about the FULCRUM PIN for 90 degrees both anticlockwise and clockwise.

3   The nut and washer moved away from the pin.

4   The pin and bush moved away from the body.

5   The arm moved vertically upwards away from the body.

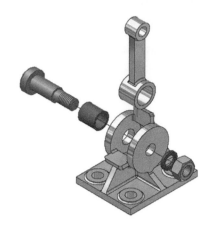

Exercise L10: shackle (H12)

1   Manual explosion with trails.

2   Ring rotated about its own vertical axis for one revolution.

3   Other parts moved in various directions for a variety of distances.

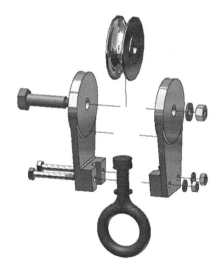

Final thoughts on this chapter

1   While the presentations created in this chapter are relatively simple, I hope that the reader understands the powerful 'tool' at their command.

2   The various projects at the end of the book will demonstrate the use of presentations and animations.

3   Saved Presentation and Animation files have different file sizes. For the L10 Shackle exercise, my file sizes were: Presentation .ipn file 139 KB, Animation Video Clip 5823 KB.

M   Adding welds

Note: Welding is a specialised topic, and those exercises in this chapter involving welding are not intended to represent usable or real-life examples of usable designs, but are provided as design exercises and for the purpose of student learning and practice only.

1   Inventor® allows welding operations on *assembled models*, and has the following weld features:
    (a)   preparation;
    (b)   welds; and
    (c)   machining.

2   For the welding exercises in this chapter, we will use previously created and saved assemblies, as well as creating new components and assemblies.

3   The procedure is:
    (a)   Open an existing .iam assembly drawing.
    (b)   Select Applications-Weldment from the menu bar.
    (c)   Use the Convert to Weldment to:
        •   set the standard (e.g. ISO); and
        •   select Welded Steel Mild – or other appropriate.
    (d)   Apply the fillet weld as appropriate.
    (e)   Save the completed work.

4   See the companion website (www.routledge.com/cw/mcfarlane) for colour versions of the models found in this chapter.

Terminology

Weld bead
An assembly feature.

Weldment
Consists of specific lengths of stock material and detailed assembled models.

Weld symbol
Gives complete information about the weld to be applied.

Weld prep

Preparing work pieces for welding by removing material along edges where the weld is to be applied. Weld prep can be:

1   by extrusion;

2   by hole; and

3   by chamfer.

Weld types

There are two types of weld available:

1   cosmetic; and

2   fillet.

Note: Welding is a specialised topic on its own, and I have no in-depth knowledge of the topic. I have included a few examples of adding fillet welds to assembled models, and I apologise to the readers if the exercises are not quite correct. Basically, I will be adding fillet welds to adjoining surfaces, which will then display the welding symbol (which will probably not be correct, but at least I tried).

Exercise M1: corner plates

1   Open the standard metric (mm) .ipt file.

2   Create a $100 \times 100 \times 10$ plate part model and save.

3   Open the standard metric (mm) .iam file.

4   (a)   Place three saved .ipt plates into the drawing.
    (b)   Assemble the three plates and 'tidy up' as displayed.

5   Add fillet welds to the plates as indicated, positioning the arrow symbols to suit.

6   Save if required.

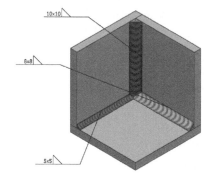

Exercise M2: angle bracket (H1)

1   Open the angle bracket assembly from Exercise H1.

2   Change to Weldment and select standards.

3   Apply suitable fillet welds between the pins and the plate, as displayed.

4   Position the welding symbol more favourably.

5   Save work.

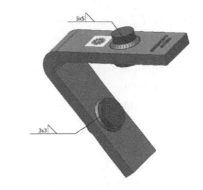

Exercise M3: back panel (H3)

1   Open the assembly from Exercise H3.

2   Change to Weldment and select standards.

3   Rotate the assembly to a suitable position and zoom in as displayed.

4   Five 4×4 fillet welds of length 10 have to be applied as indicated.

5   Position the welding symbol more favourably.

6   Save work if required.

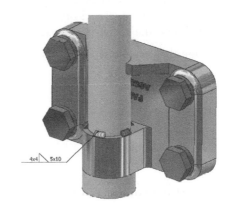

Exercise M4: point marking device (H6)

1   Open assembly from Exercise H6.

2   Change to Weldment and select standards.

3   Rotate the assembly to a suitable position and zoom in as indicated.

4   Apply a suitable fillet weld, as displayed.

5   Add a cosmetic weld where indicated.

6   Position the welding symbol more favourably.

7   Save work if required.

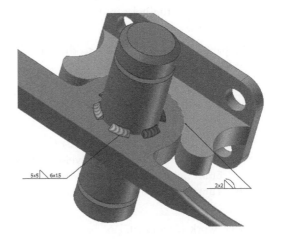

Exercise M5: steam venting valve (H7)

1   Open the Exercise H7 assembly, which requires the following modifications:
    (a) remove/delete the four square headed bolts from the top flange; and
    (b) plug the four holes in the top flange (as displayed).

2   Change to Weldment and select standards.

3   Rotate the assembly to a suitable position and zoom in as indicated.

4   Apply a 2 × 2mm chamfer weld preparation to the top edge of the valve body and to the lower edge of the top flange.

5   Add a 2mm fillet weld to the appropriate faces (note that I have displayed only a partial fillet weld around the chamfer preparation).

6   Position the welding symbols more favourably.

7   Save work if required.

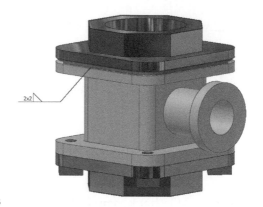

Exercise M6: process adaptor (H8)

1   Open the process adaptor assembly from Exercise H8.

2   Change to Weldment and select standards.

3   From the assembly, delete the four M10 bolts and 'remove' the holes and hole/bolt constraints.

4   Prepare a 5 × 5mm chamfer to the top edge of the BASE PLATE – part 1.

5   Add 12 5 × 5mm intermittent welds of length 10mm to the chamfer region, positioning the welding symbol to suit.

6   Add the following cosmetic welds:
    (a) 10 2 × 2mm intermittent welds of length 5mm between the body and the cap; and
    (b) 1mm V–Butt between the cap and the cap top.

7   Save work if required.

Exercise M7: roller spindle (H9)

1 Open the Exercise H9 assembly.

2 Change to Weldment and select standards.

3 Remove from the assembly:
(a) the M12 nut and washer; and
(b) the two thread displays.

4 Modify the length of the (part 3) ROD from 153mm to 145mm.

5 Prepare chamfer to two edges of the FRAME, as displayed.

6 To the two prepared chamfer regions, add the appropriate fillet welds, as displayed, and position the welding symbol as required.

7 Save work if required.

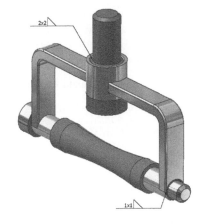

Exercise M8: tool holder (K11)

1 Open the assembly in Exercise K11.

2 Change to Weldment and select standards.

3 Remove the M10 nuts and the thread display from the assembly.

4 Add a 5mm fillet weld between the top annular surface of the handle and the (now) unthreaded rod.

5 With the Machining weld feature, add a 5mm hole between the top of the added fillet weld and the rod (some thought needed for this hole to be added).

6 Save work if required.

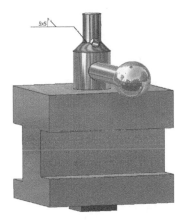

Exercise M9: swinging arm rig (H10)

1 Open the swinging arm rig assembly in Exercise H10.

2 Change to Weldment and select standards.

3 Remove the four M14 bolts and their 'associated' holes and constraints from the appropriate parts of the assembly.

4 Add suitable fillet welds between parts 1 and 2 (BASE and UPRIGHT).

5 My display illustrates two possibilities, but I have removed the weld symbol, so you will need to 'design' your own weld.

6 Save work if required.

Exercise M10: shackle (H12)

1   For the shackle assembly:
   (a) change to Weldment and select standards; and
   (b) remove the bottom two bolts, washers, nuts, holes and
       constraints.

2   Add a cosmetic weld of your choice between the two BODY parts.

3   Save work if required.

4   *Question*: Can you chamfer the 'split' between the two BODY parts
and add a fillet weld?

Exercise M11: inspection cover (J1)

The inspection cover to be used in this exercise requires another part, so that an assembly can
be created, so:

1   Using the drawing information below, create a base for the inspection cover, then assemble
the actual inspection cover onto this base with suitable constraints.

2   Remove all holes from the cover.

3   Add suitable welds (cosmetic and/or fillet) between the new base and the cover flanges,
similar to the enlarged display below.

4   Repeat 'at the other end', then save if required.

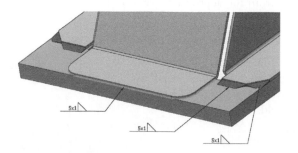

Exercise M12: pipe weld

1   Two pipes have an OD of 64mm and are 10mm thick.

2   The pipe lengths are 192mm and 114mm, and they require some 'cuts' prior to being assembled.

3   Using the drawing information below, create the two pipes, then create the assembly, which will also be used in the next two exercises, so additional 'saves' are required.

4   The first weld exercise requires a weld to be applied to the pipe join.

5   Create this weld to your own specification.

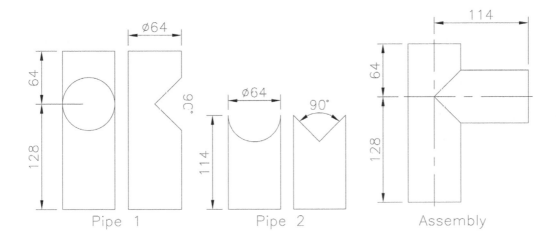

Pipe 1   Pipe 2   Assembly

Note: It had been my intention to create a 10 × 10mm fillet weld between the pipe joins, but every time I attempted it, my system crashed with the 'error log being created' message, hence the cosmetic weld. My apologies for this. Hopefully, you will be able to create a fillet weld.

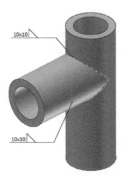

Exercise M13: pipe weld with support piece

1   The two-pipe assembly created in Exercise M12 requires a triangular strengthening piece to be added.

2   Using the drawing information below, create this additional part, then assemble it 'onto' the assembly and add suitable welds to your own specification.

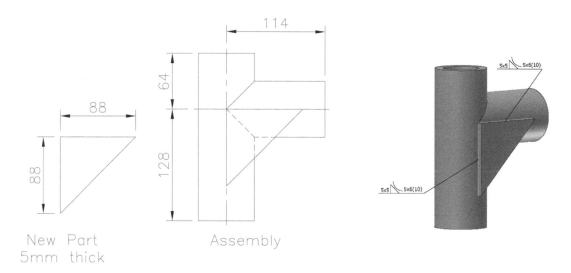

New Part
5mm thick

Assembly

Exercise M14: pipe weld with gusset support

1   The two-pipe assembly created in Exercise M12 is still not satisfactory for the design department.

2   They have suggested the addition of a gusset piece to assist with various stress calculations.

3   Using the drawing information below, create this gusset piece, then assemble it 'onto' the assembly and add suitable welds to your own specification.

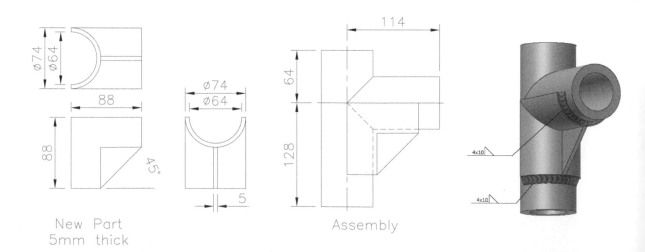

New Part
5mm thick

Assembly

Exercise M15: 'T' plate assembly

1   Two steel plates have to be welded together to form a 'T' shape.

2   The weld information is as follows:
    (a) staggered intermittent fillet with 'depth' of 8mm;
    (b) four on the near side at a length of 20mm with gaps of 10mm; and
    (c) the same on the far side.

3   The plate details are:
    (a) horizontal plate is $110 \times 60 \times 10$; and
    (b) vertical plate is $110 \times 60 \times 20$.

4   Create an 'assembly' of the two plates with the vertical plate 'central' to the horizontal plate.

5   Add the staggered intermittent fillet weld using the information given.

6   'Insert' the appropriate weld symbol.

7   When the assembly with the welds added is complete, create a First Angle drawing layout.

8   This exercise is courtesy of Debbie McMahon, CAD apprentice technician at Doosan Babcock, Renfrew. She informed me that the weld symbol was 'not quite correct'.

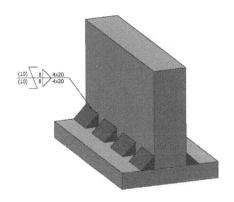

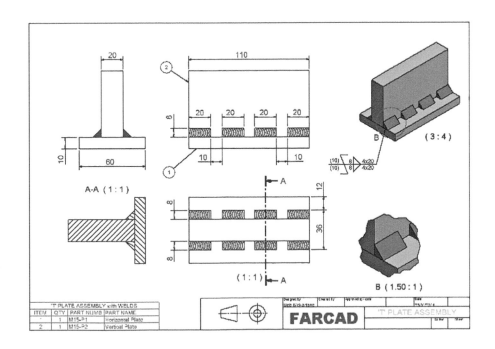

N Applying your skills

1   This chapter will cover all the topics from previous chapters (i.e. part models, assemblies, sheet metal, drawing layouts, presentations and welding).

2   The reader should decide what Inventor® 'sheet' to start with and how to complete the various exercises.

3   The information given will be (as usual) orthographic drawings with a minimum of text.

4   Hopefully, these exercises will prove worthwhile and enjoyable, and they will certainly 'test' your knowledge and skills.

5   As with previous chapters, there is no order to the exercises; they are placed as I completed them.

6   Remember:
    (a) save all work to a suitably named folder; and
    (b) discretion for dimensions, etc. which are unclear.

7   Any animations that I have created can be accessed on the companion website (www.routledge.com/cw/mcfarlane), in the folder of sample files for animations. Again, this folder will also contain some colour versions of the models in this chapter.

Exercise N1: bench vice

1   A bench vice consists of a Base, Fixed Jaw, Movable Jaw, Screw Thread and Toggle, and is displayed with dimensions below.

2   Create part models of each part, then create:
   (a)   a bench vice assembly; and
   (b)   a First Angle projection drawing layout with a parts list, the views being to your specification.

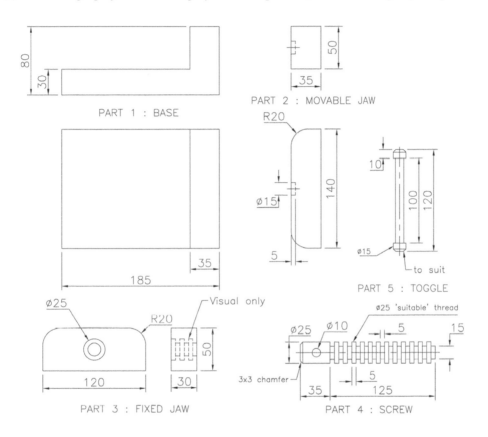

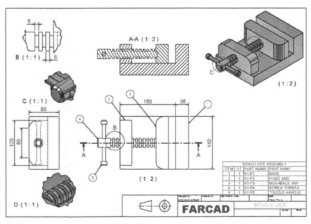

Exercise N2: steady

Use the two orthogonal First Angle projection views (with an auxiliary) to create the part model.

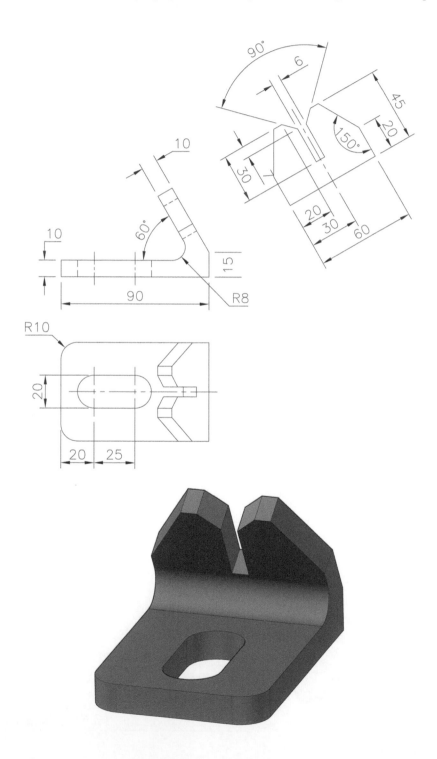

Exercise N3: rod end

Use the information given in the two views to create a model of the rod end.

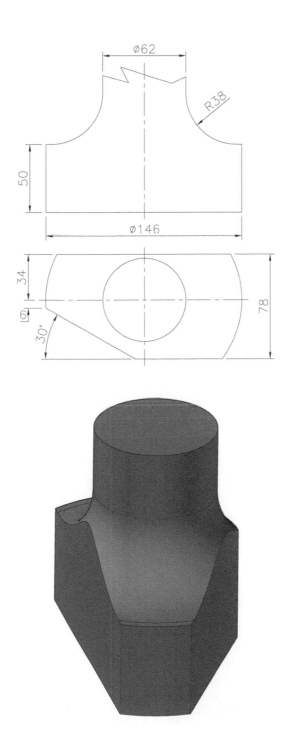

Exercise N4: casting

Create a model for the casting using the dimensions given.

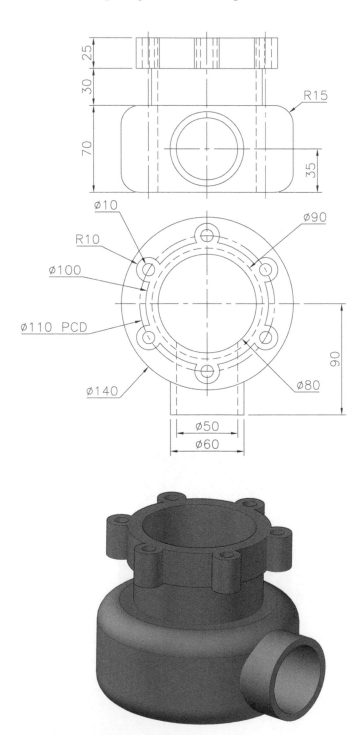

Exercise N5: support 1

Create the model of the support using the drawing information given.

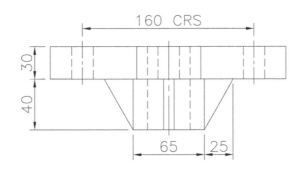

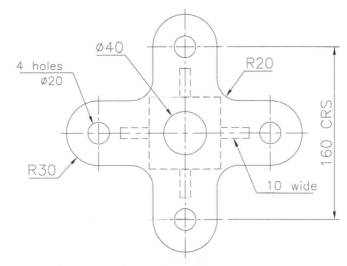

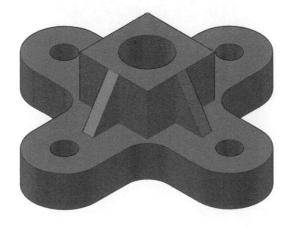

Exercise N6: support 2

Use the drawing information below to create a model of the support.

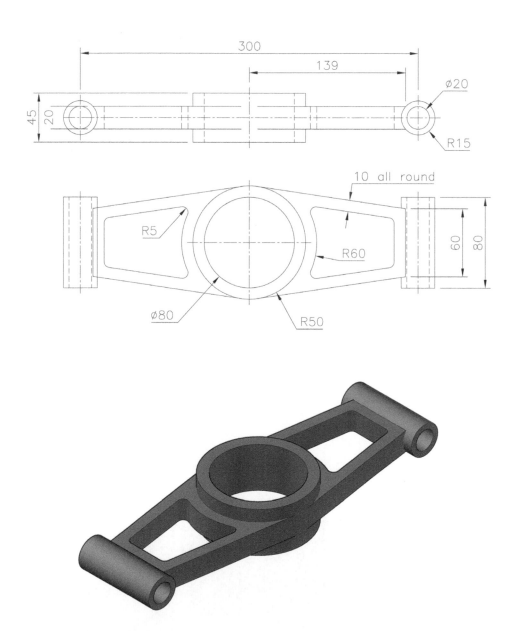

Exercise N7: gimlet

Use the drawing information to create a model of the gimlet and then create a drawing layout to your specification.

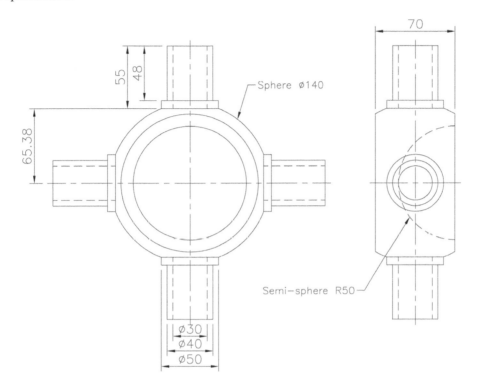

Sphere ∅140

Semi–sphere R50

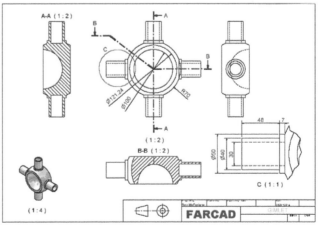

Exercise N8: chain

1 The first 'link' in a chain has been created from a Ø20 circle 'swept' around a Ø100 circle.

2 The second 'link' in the chain has the same 'shape' as the first link, but is 'turned through' 180 degrees.

3 Create a straight length of this chain to display (at least) 10 links.

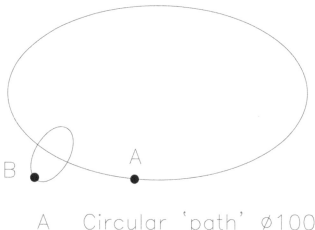

A   Circular 'path' Ø100
B   Circle Ø20

Exercise N9: faceplate

Using the information in the display below, create a model display of the faceplates, which are 100mm apart.

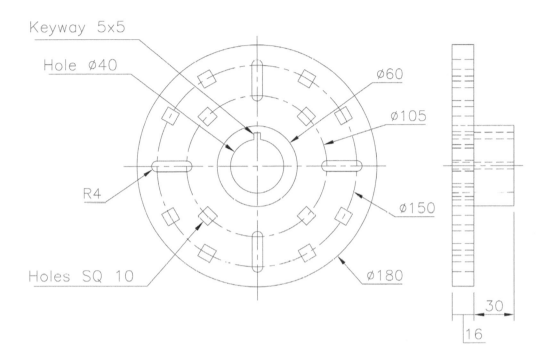

Exercise N10: lathe milling wheel holder assembly

Note: This exercise has additional tasks after step 2.

1   The parts for this assembly are (part no: part name (no of)):
   (a) Part 1: T Bar (1).
   (b) Part 2: Arms (2).
   (c) Part 3: Washer (2).
   (d) Part 4: Screw (1).
   (e) Part 5: Nut (2).
   (f) Part 6: Pin (2).

2   Use the drawing information below to create an assembly and drawing layout of the holder.

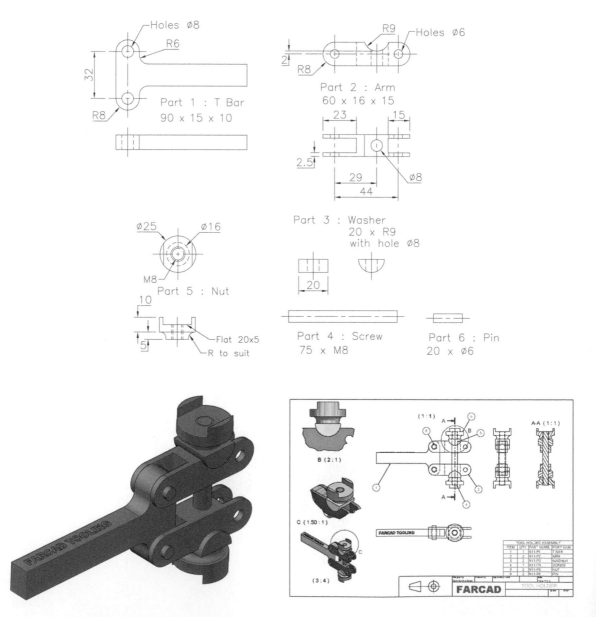

3 Modification to the assembly:
   (a) Due to a possible 'design fault', it was decided to modify the original design by adding a bead of weld between parts 2 and 3 (the arms and the washers).
   (b) This was achieved with:
   * preparation: a 3 × 3 chamfer on the curved face of the part 2 arms;
   * weld: a 4mm fillet weld 'added' between this chamfer and the curved surface of part 3, the washers; and
   * machining: a 3mm diameter hole added to the lower quadrant of the part 3 washer.

4 Complete the design changes, and if saving these changes, take care with the file name.

5 Use the parts list to create a spreadsheet (to your own specification) with additional data added, similar to that displayed below.

ITEM	QTY	PART NUMBER	PART NAME	UNIT COST (£)	No req	Total Cost (£)
1	1	N11-P1	T BAR	1.1	500	550
2	2	N11-P2	ARM	0.95	1000	950
3	2	N11-P3	WASHER	0.85	1000	850
4	1	N11-P4	SCREW	0.3	750	225
5	2	N11-P5	NUT	0.49	1890	926.1
6	2	N11-P6	PIN	0.08	3000	240
				Customer Manuf Cost:		3741.1

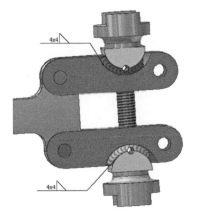

6 Exploded view and animation: Create a manual exploded view and animation of the assembly with your own entries.

Exercise N11: stand and canopy

Create a 'sectioned' two-part assembly using the information provided.

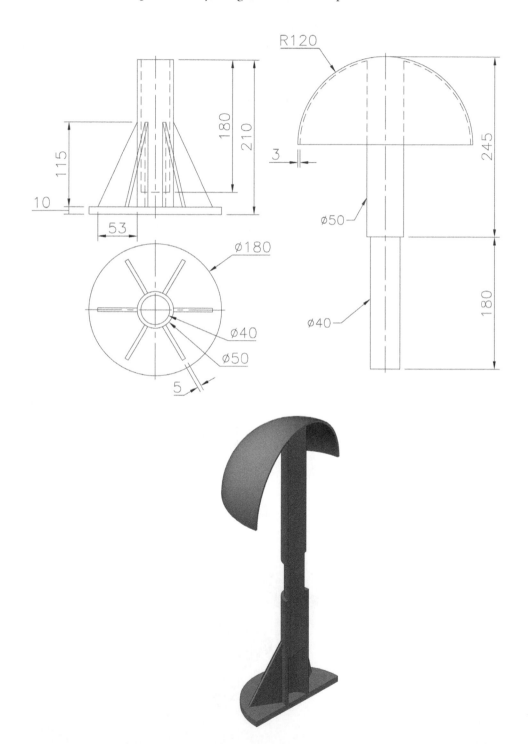

Exercise N12: engine casting

Create a model and drawing layout of the casting using the information displayed.

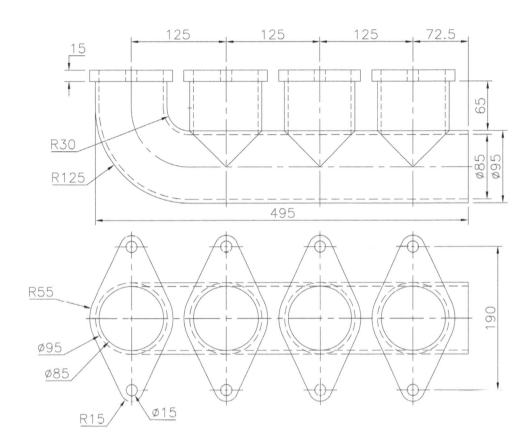

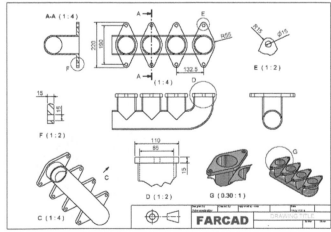

Exercise N13: car wheel

Create a model of the car wheel using the information displayed (or design your own).

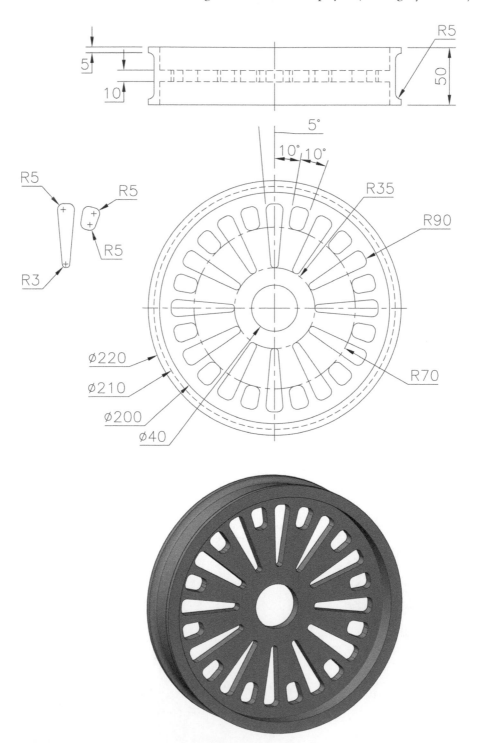

Exercise N14: panels

1 Two panel models to be created using the drawing details provided.

2 Note that every 'horizontal section' is a regular octagon/hexagon, so these models should be relatively easy.

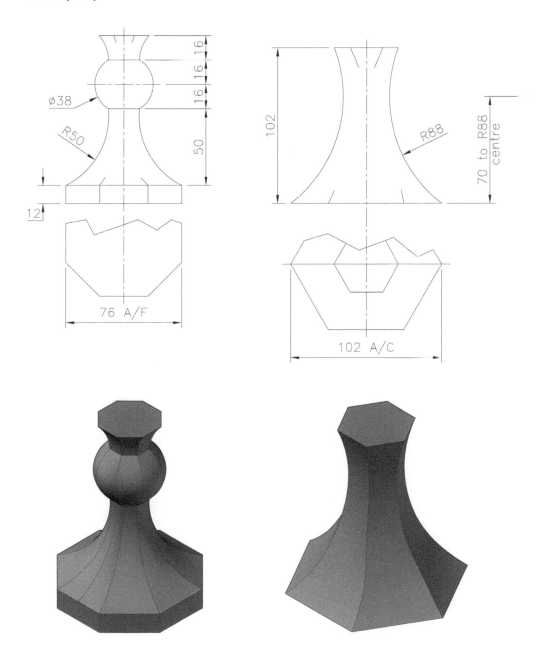

Exercise N15: steam valve body modification

1   Several parts of the steam valve body assembly (Exercise H7) have been modified and a new assembly is required.

2   Do you work with the original assembly, the original Exercise H7 parts or create new parts? Your decision.

3   The modifications are as follows:
    (a) *Assembly*
        •   all the square head bolts have been removed; and
        •   the bottom flange has been removed.
    (b) *Body*
        •   all the holes have been removed;
        •   the lower face of the body has had its 'depth' increased from 10 to 20mm; and
        •   the central hole 'depth' is now 60mm.
    (c) *Flange*
        1   the four holes have been removed.
    (d) *Gasket*
        •   a 1.6mm sheet metal copper gasket has been introduced between the top flange and the valve body; and
        •   the gasket has the 20mm rounded corners and the 70mm central hole.

4   Create the new assembly, and display as an exploded view.

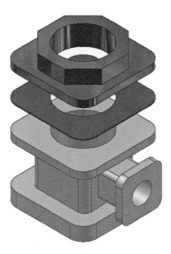

Exercise N16: 90–degree bracket

Create a Third Angle projection drawing layout using the drawing information below.

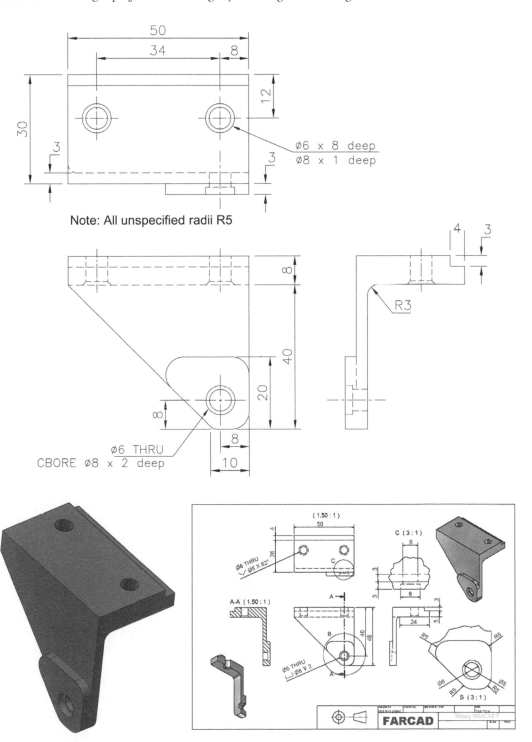

Exercise N17: shaft box

Create an exploded view, a First Angle drawing layout and an animation of the assembly.

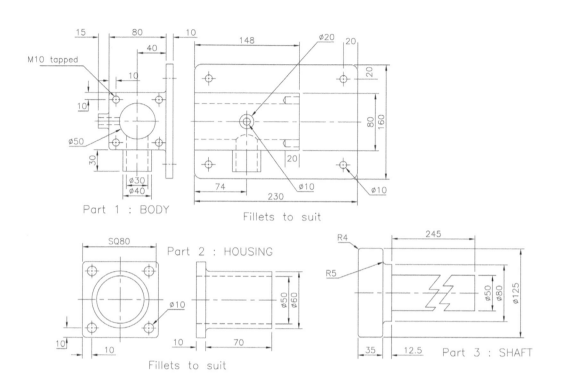

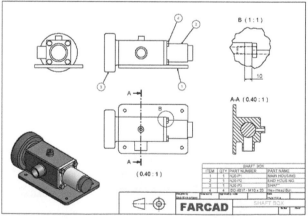

Exercise N18: cubic space division

(with acknowledgement to M. C. Escher)

1 The work of M. C. Escher is well known.

2 He produced amazing graphical work and optical illusions before computers.

3 Some of his most famous works include: 'The Waterfall', 'Ascending and Descending Stairs' and 'Metamorphosis', as well as 'Cubic Space Division'.

4 The model to be created in this exercise is based on 'Cubic Space Division'.

5 The 'basic shape' for the model is a 50mm cube with cylinders of ⌀25mm and length 100mm.

6 You have to design a $10 \times 10 \times 10$ rectangular pattern using this basic information and display at a suitable viewpoint.

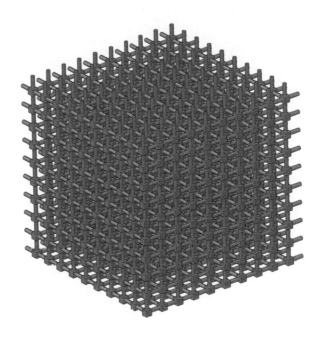

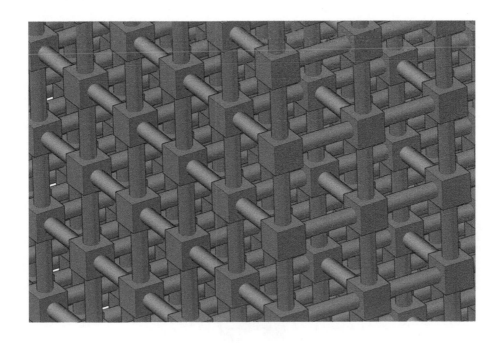

Exercise N19: wing support bracket

Using the information in the display below, create a model of the component.

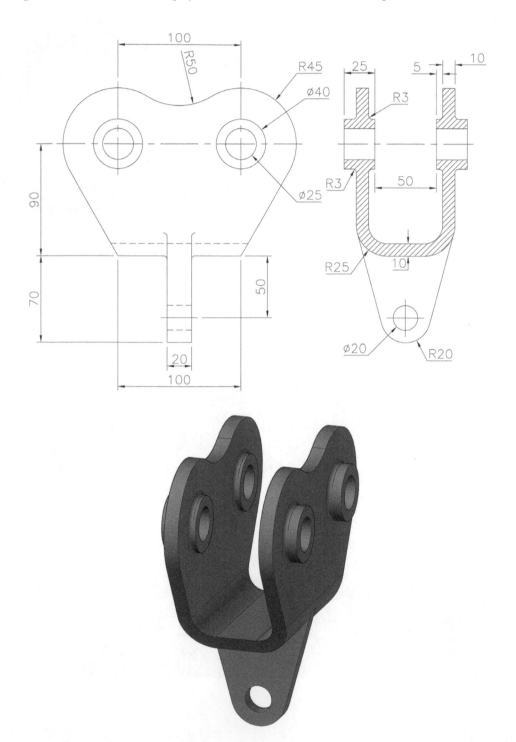

Exercise N20: offset wrench

Use the drawing information below (with the dimensions in inches) to create a model of the component, which is 1 inch wide.

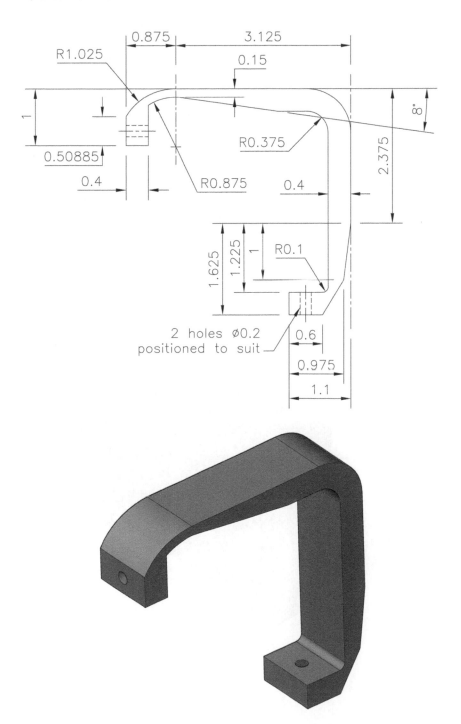

Exercise N21: bearing base

Create a model of the bearing base using the Third Angle drawing displayed.

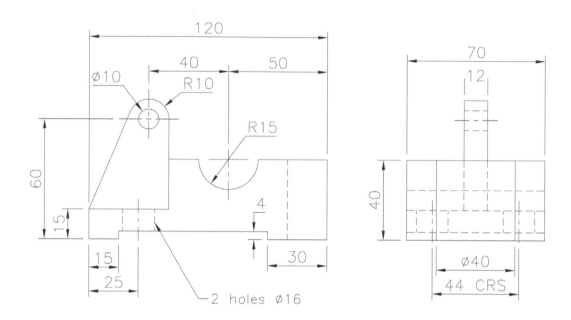

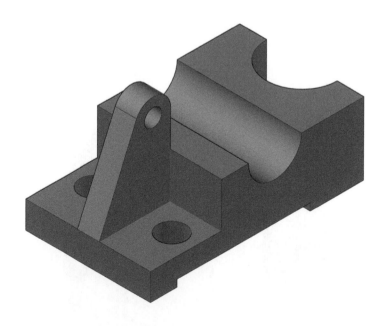

Exercise N22: pulley

Use the drawing information to create and display the model to display the features.

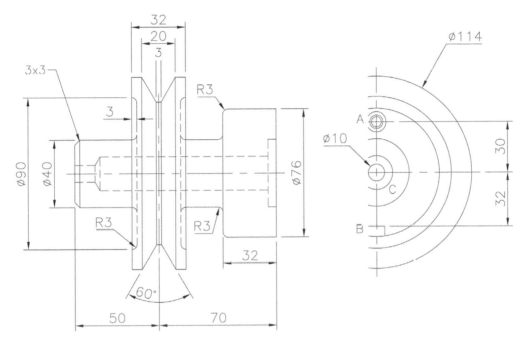

A: Tap M8−6H
   15 min length full thread
   C'BORE ∅12 x 3 deep

B: SLOT 10 wide
   22 long

C: DRILL ∅20 x 100 deep
   C'BORE ∅10 x 5 deep

Exercise N23: cover

A front elevation and two end elevations in First Angle projection from which to create a First Angle drawing layout (to your own specification) of the cover model.

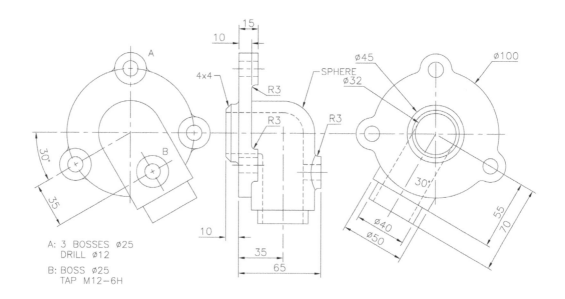

A: 3 BOSSES ⌀25
   DRILL ⌀12

B: BOSS ⌀25
   TAP M12–6H

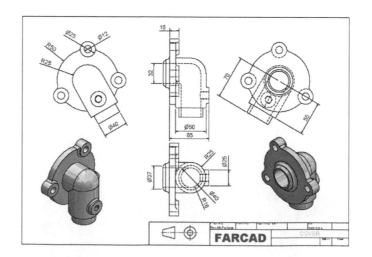

Exercise N24: Watt's straight line motion

1   This exercise is completely different from any previously attempted.

2   Most engineers will know about Watt's straight line motion, based on linkages.

3   In this exercise, we will 'attempt to animate' this straight line motion using a simple assembly.

4   (a)   There are four parts in the assembly, these being lever 1 (2 of), lever 2 (1 of), pin (4 of) and an anchor point (2 of).

    (b)   You are required to design your own anchor.

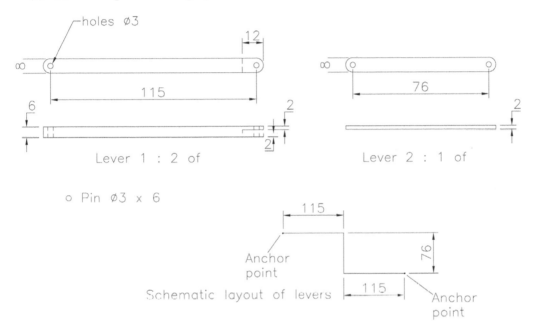

Lever 1 : 2 of   Lever 2 : 1 of

o Pin Ø3 x 6

Schematic layout of levers

5   The levers are positioned as displayed, with pins completing the assembly.

6   The anchor points are required to 'ground' the assembly and only allow the levers to rotate about their pin ends.

7   Now animate the assembly and prove Watt's straight line motion for the 'middle position' of lever 2.

8   I have displayed my animated assembly with screen dumps for four different linkage positions, and there is a straight line between the lever 2 midpoints.

9   As I said, this has been a totally different type of exercise, which I hope you found interesting and worthwhile.

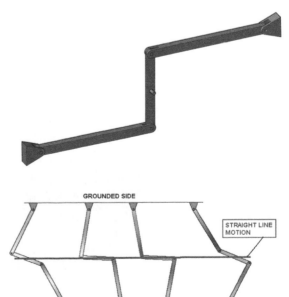

Exercise N25: nickel casting

Use the drawing information to create the model.

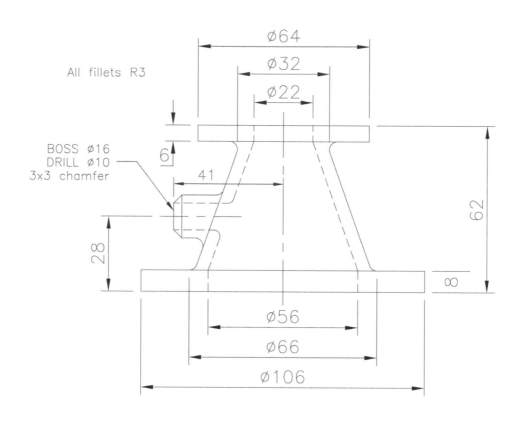

Exercise N26: jig detail

Use the drawing information to create the model.

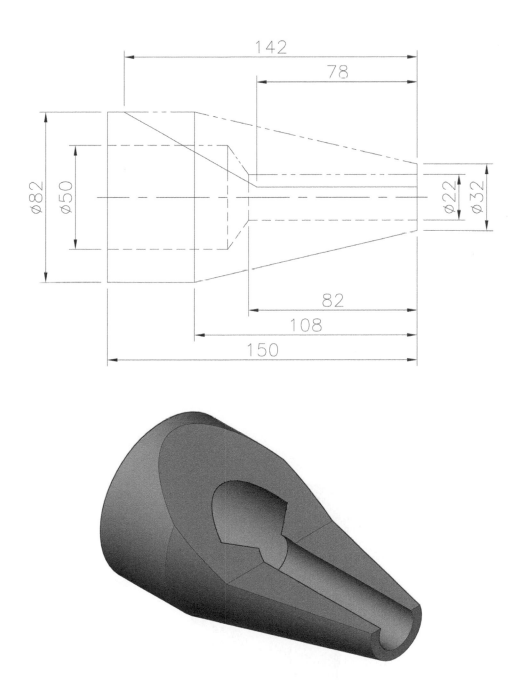

Exercise N27: wing nut

Create a model of the wing nut using the drawing information below.

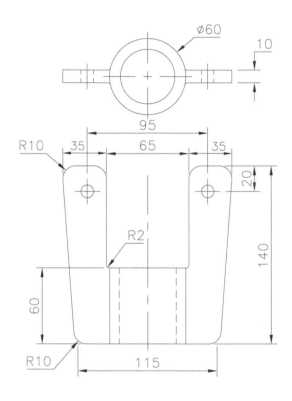

Exercise N28: conical chute

1   A conical chute is 'arranged' around a 96mm cone, details of which are displayed below.

2   Create a two-revolution chute using the information given.

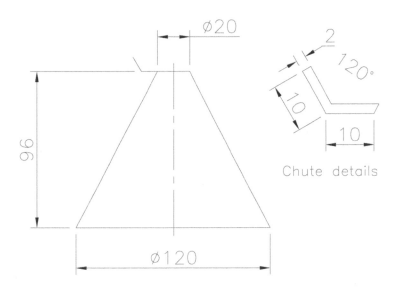

Chute details

Exercise N29: sheet metal jug

1   The body of a metal jug is constructed from two cones, arranged as below.

2   Sheet aluminium of 2mm thickness is used.

3   Create a model of the jug using the information given.

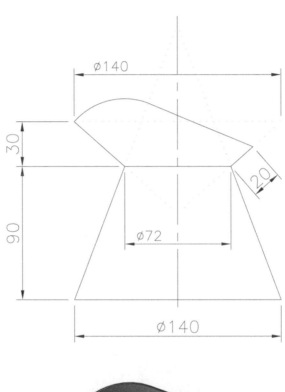

Exercise N30: clamping arrangement

1 The parts for this assembly are detailed below.

2 The clamp (3) is secured to the 250×150 base plate (1) by an M12 stud (4) with a hexagonal nut and washer.

3 The elliptical flange (2) is located on the 50mm diameter spigot on the base plate.

4 Create an assembly of the model, then create a drawing layout to your own specification, with parts identified with a parts list.

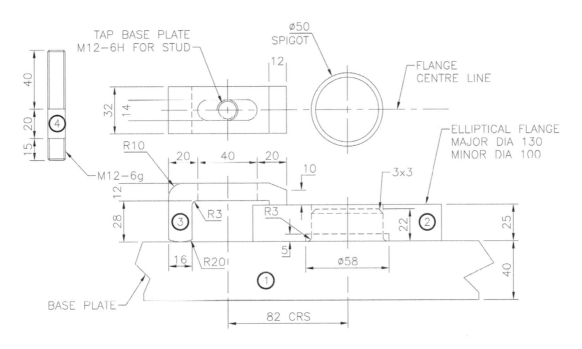

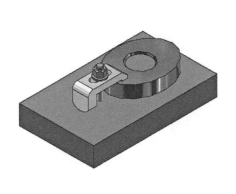

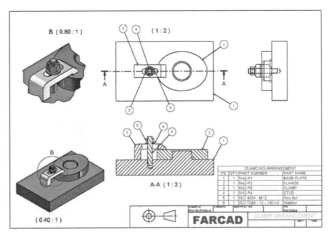

Exercise N31: small pipe support

1   The parts for this assembly are detailed below.

2   The cap (1) is secured to the body (2) by an M6 stud (3) with hexagon nut and washer on one side and by an M6 hexagonal bolt with nut and washers on the other side.

3   Create an assembly of the model.

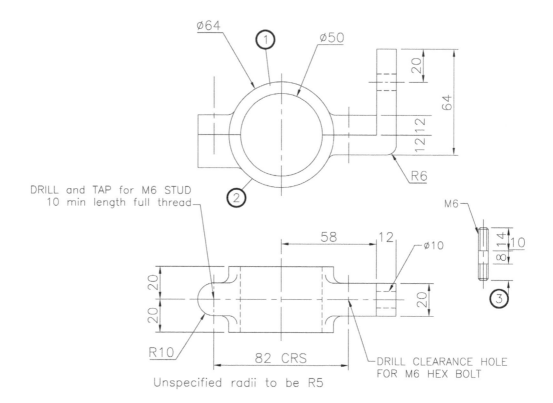

DRILL and TAP for M6 STUD
10 min length full thread

R10

82 CRS

Unspecified radii to be R5

DRILL CLEARANCE HOLE
FOR M6 HEX BOLT

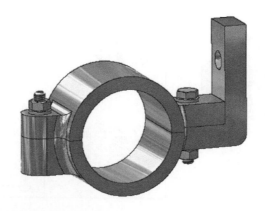

Exercise N32: four hopper chutes

1   Using the drawing information displayed, create models of the four hopper chutes.

2   The material used to create the chutes is 1.5mm galvanised metal.

Chute 1

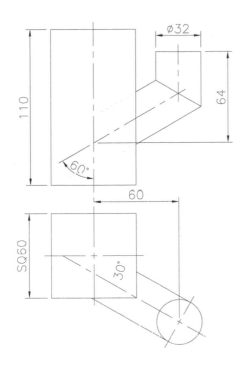

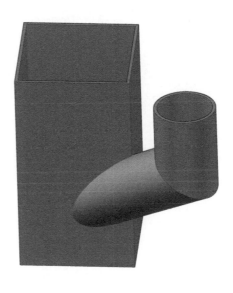

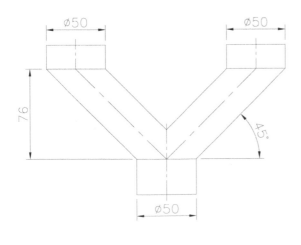

Chute 2

Chute 3

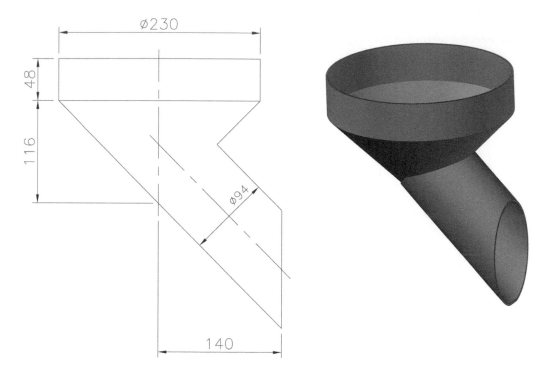

Chute 4

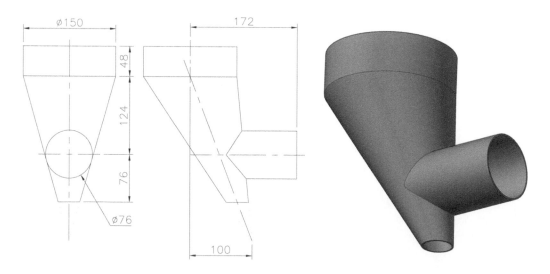

Exercise N33: two vent cowls

1   Using the drawing information displayed, create models of the two cowls.

2   The material used to create the cowls is 2mm aluminium metal.

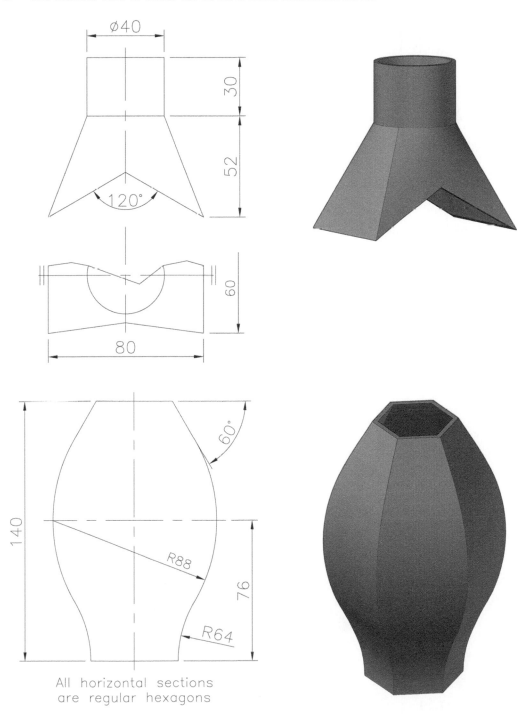

Ø40

30

52

120°

60

80

60°

140

R88

76

R64

All horizontal sections
are regular hexagons

Exercise N34: moulded column base

Create a model of the column base using the drawing information given.

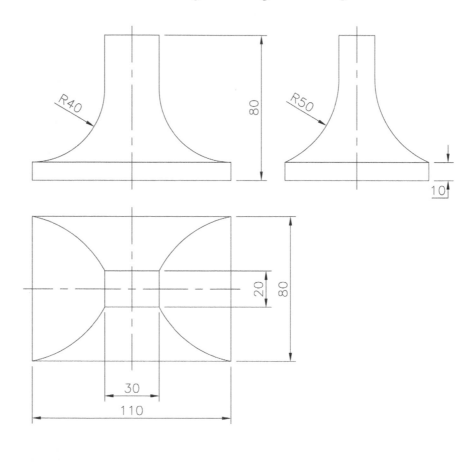

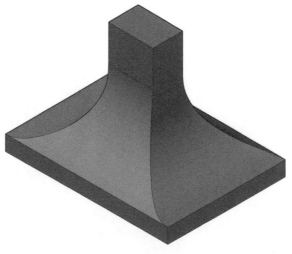

Exercise N35: plastic garden tool handle

Create the model using the drawing information given.

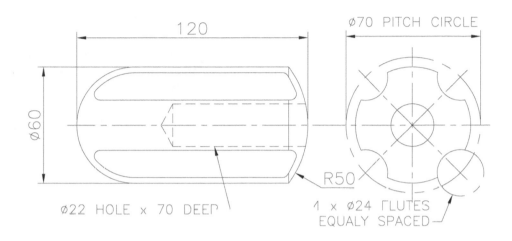

120

∅60

∅22 HOLE x 70 DEEP

∅70 PITCH CIRCLE

R50

1 x ∅24 FLUTES
EQUALY SPACED

Exercise N36: two sheet metal models

Model A

Using the drawing details below, create a model from 1.6mm brass sheet. There are two fold lines, as indicated.

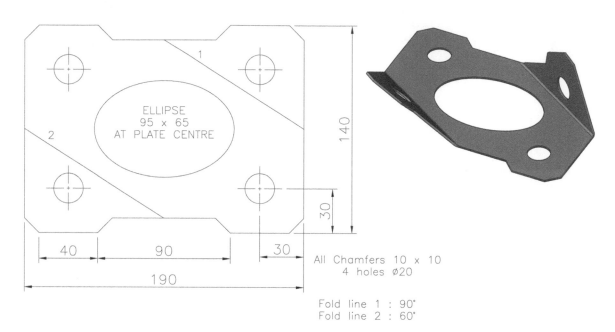

All Chamfers 10 x 10
4 holes ⌀20

Fold line 1 : 90°
Fold line 2 : 60°

Model B

Using the drawing details below, create a model from 1.6mm copper sheet with a single fold line (at 42° to the horizontal), as indicated.

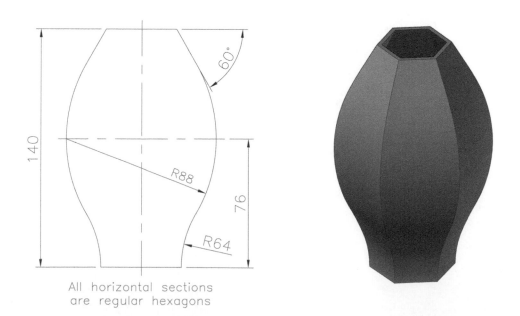

All horizontal sections
are regular hexagons

Exercise N37: fitting

Create the model from the drawing information displayed.

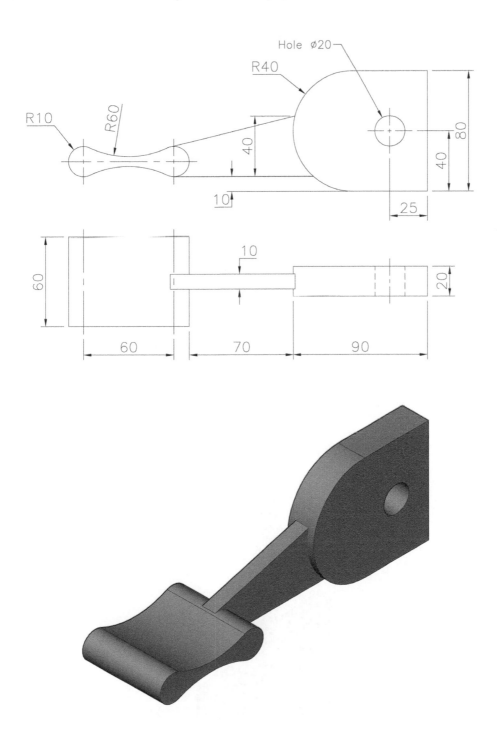

Exercise N38: plastic impellor

Create the model from the drawing below.

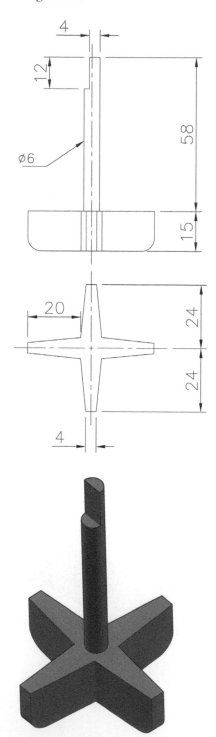

Exercise N39: pump cover

Use the drawing information (and your discretion) to create a model of the pump cover.

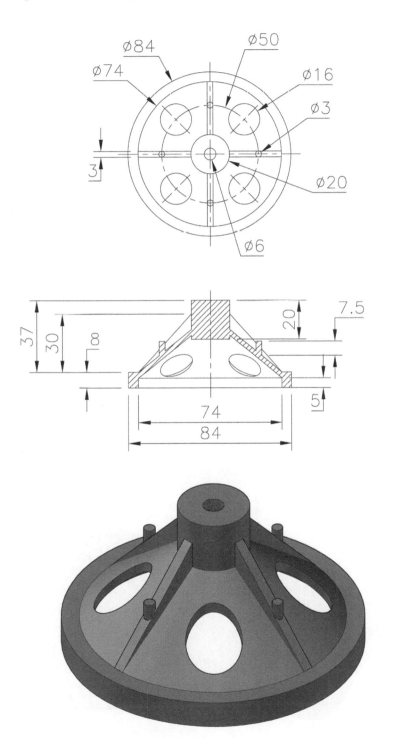

Exercise N40: tidy

Create a model of the tidy using the drawings displayed. The thickness is 5mm.

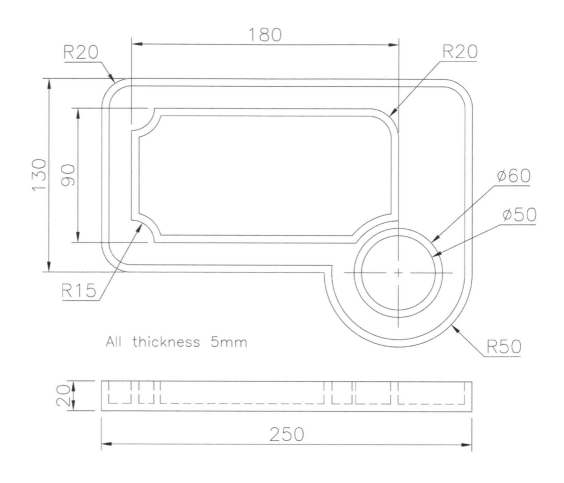

All thickness 5mm

O Reading engineering drawings

1   A few 'traditional' type engineering drawings, which require to be read and from which a model has to be created.

2   Additional drawing layouts are at your discretion.

3   The First and Third Angle projection views were created using AutoCAD, so if there are any issues with wrong/omitted dimensions, I apologise in advance, but you should have the ability to determine any 'wrong' dimensions (i.e. use your discretion).

4   As the section is basically reading engineering type drawings, there will be references to webs, bosses, fillets (which can be a 'nuisance'), A/F, etc., which should not give the reader any problems.

5   As with previous sections, there is no order of difficulty to the exercises.

Exercise O1: pulley wheel

1   A simple exercise to start the chapter.

2   Create a model of the pulley wheel using the dimensions given in the First Angle projection drawings.

3   Use your discretion with the 'five cut-outs', as my end view has not been completed.

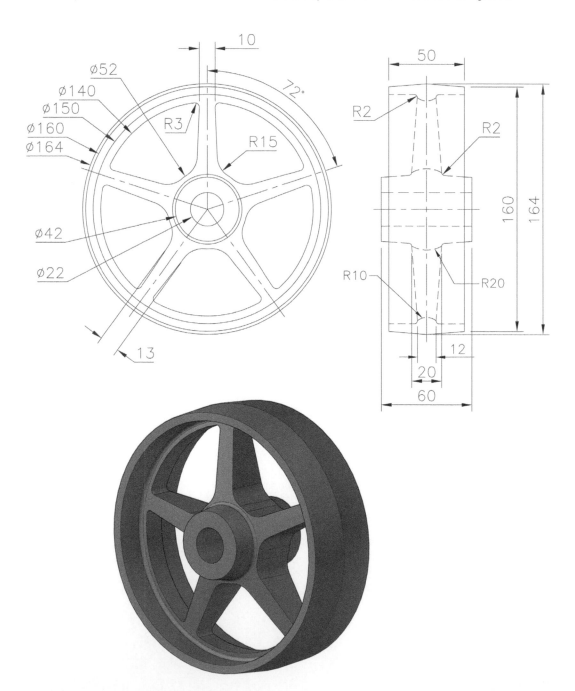

Exercise O2: camshaft bearing

Create a model of the camshaft bearing using the First Angle drawing given.

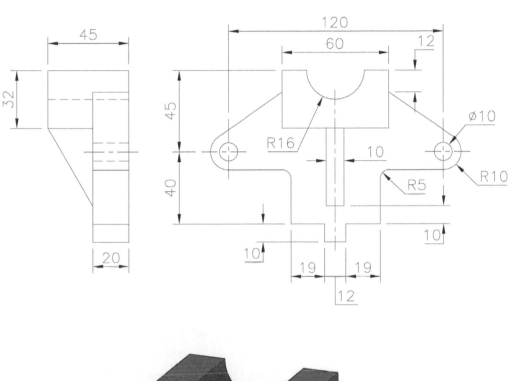

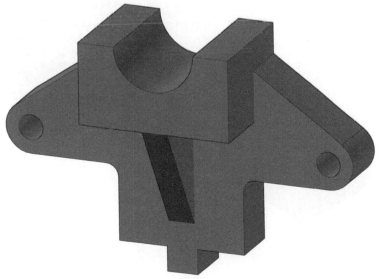

Exercise O3: clamp

Create a model of the component using the Third Angle drawing displayed.

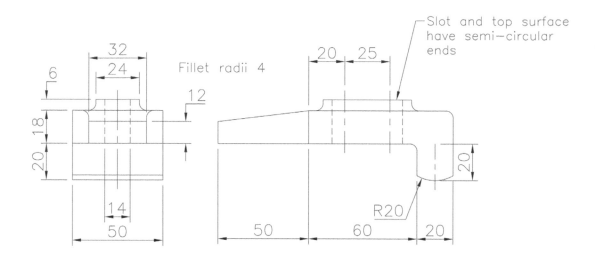

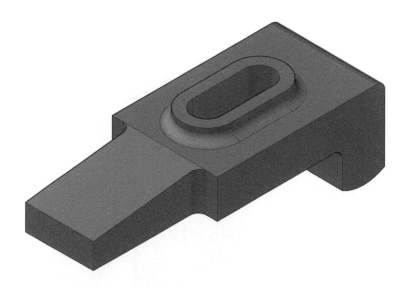

Exercise O4: angled wedge

Create a model of the component using the First Angle dimensions given.

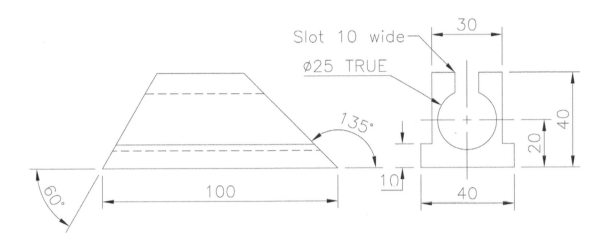

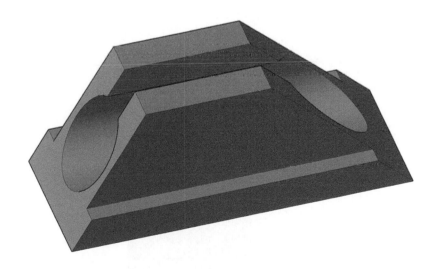

Exercise O5: metal clip

Create a model of the component using the First Angle dimensions given.

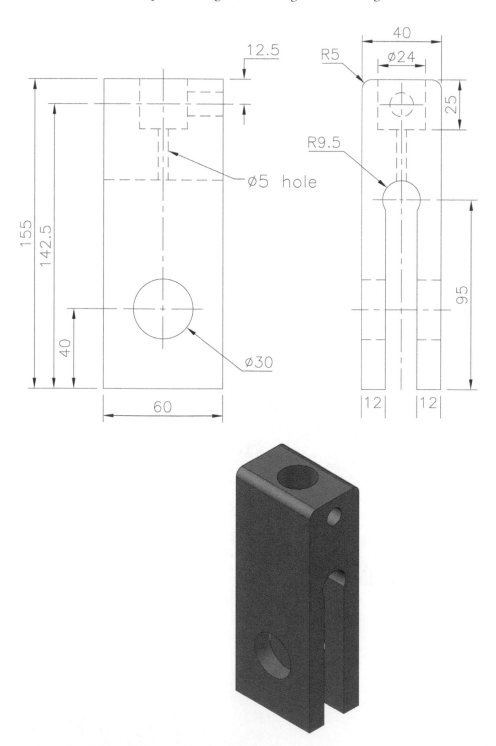

Exercise O6: bearing

Create a model of the component using the First Angle dimensions given.

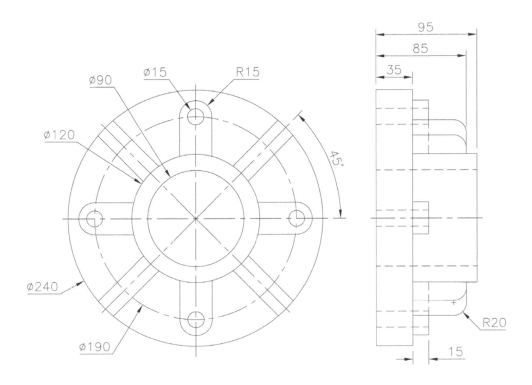

Exercise O7: expansion box

Create a model of the component using the First Angle dimensions given.

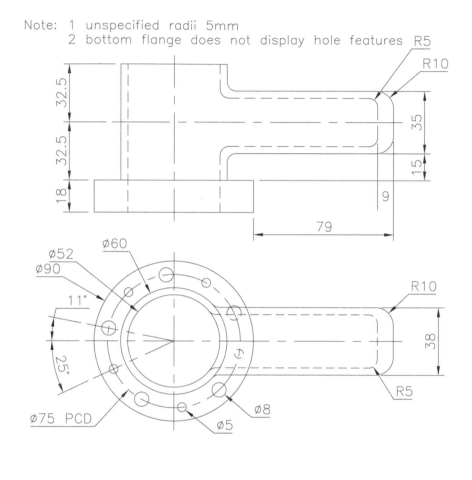

Note: 1 unspecified radii 5mm
2 bottom flange does not display hole features

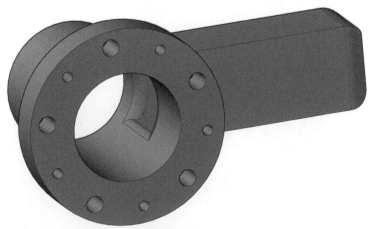

Exercise O8: base block

Using the First Angle drawing details and text information displayed, create a model of the base block.

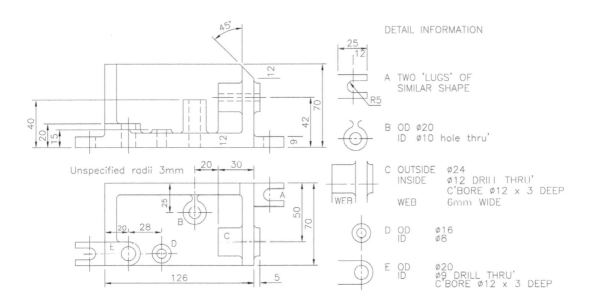

DETAIL INFORMATION

A TWO 'LUGS' OF
   SIMILAR SHAPE

B OD Ø20
   ID Ø10 hole thru'

C OUTSIDE Ø24
   INSIDE   Ø12 DRILL THRU'
             C'BORE Ø12 × 3 DEEP
   WEB   6mm WIDE

D OD   Ø16
   ID   Ø8

E OD   Ø20
   ID   Ø9 DRILL THRU'
        C'BORE Ø12 × 3 DEEP

Unspecified radii 3mm

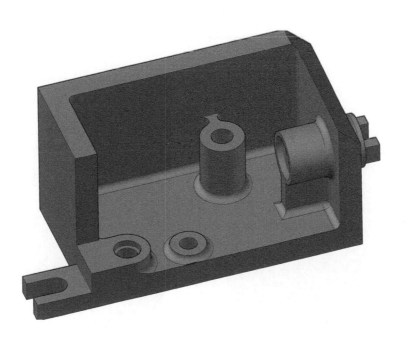

Exercise O9: crank disc

Using the Third Angle drawing and part section details, create a model of the component.

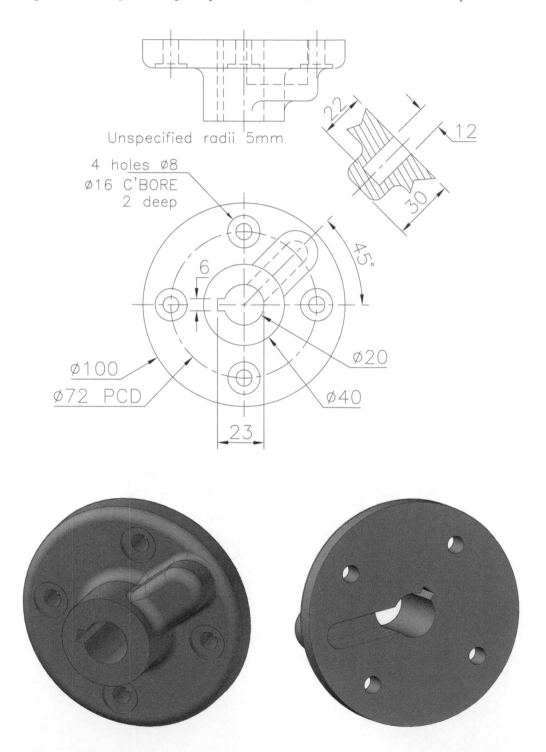

Unspecified radii 5mm

4 holes ∅8
∅16 C'BORE
2 deep

22

12

30

45°

6

∅100
∅72 PCD

∅20

∅40

23

Exercise O10: sealing cap

Using the Third Angle drawing details, create a model of the component.

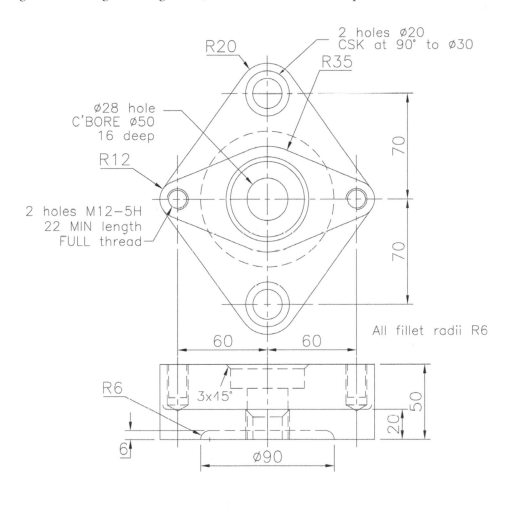

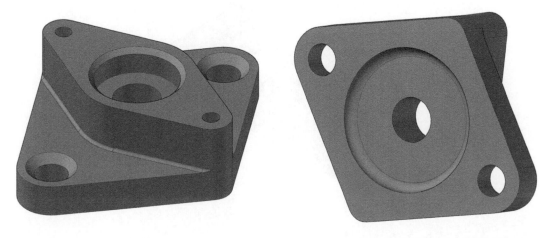

Exercise O11: clutch bracket

Use the First Angle projection drawings to create a model of the clutch bracket, using your discretion as appropriate.

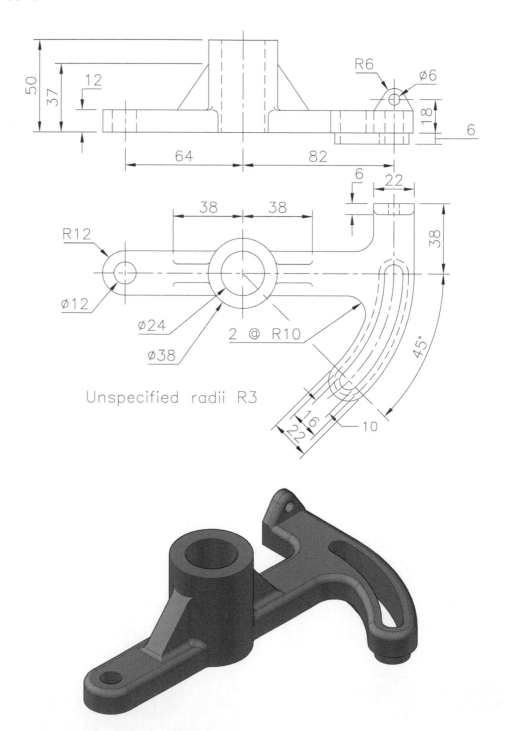

Exercise O12: spindle housing

Use the two Third Angle projection views to create a model of the component. As always, use your discretion for unsure dimensions.

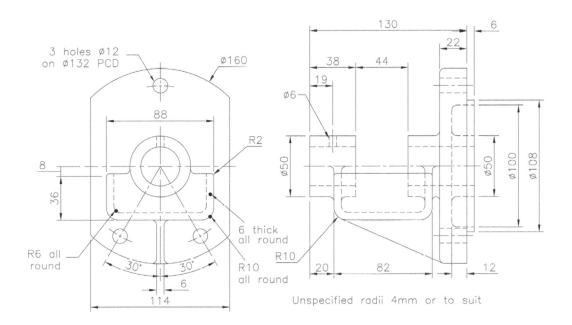

Unspecified radii 4mm or to suit

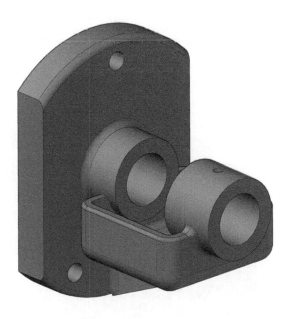

Exercise O13: intermediate casing

Two First Angle views of a casing to be modelled.

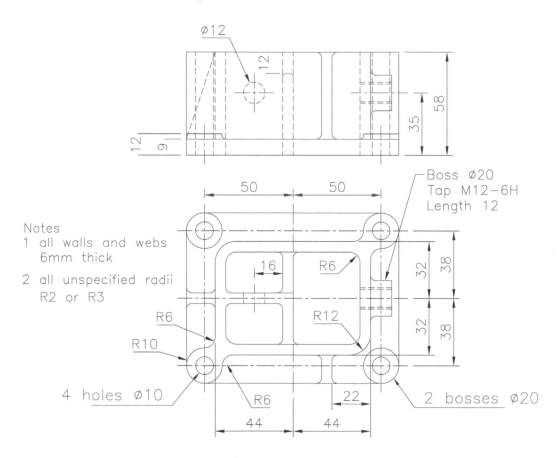

Notes
1 all walls and webs
   6mm thick

2 all unspecified radii
   R2 or R3

Boss Ø20
Tap M12–6H
Length 12

4 holes Ø10

2 bosses Ø20

Exercise O14: angle bracket

Create a model of the angle bracket from the two First Angle projection views.

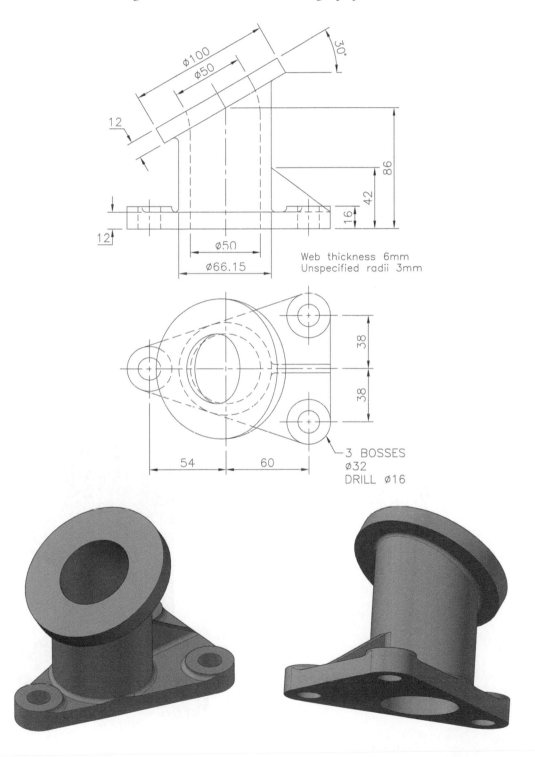

Web thickness 6mm
Unspecified radii 3mm

3 BOSSES
ø32
DRILL ø16

Exercise O15: indicator sleeve

Create a model of the component using the two First Angle projection views.

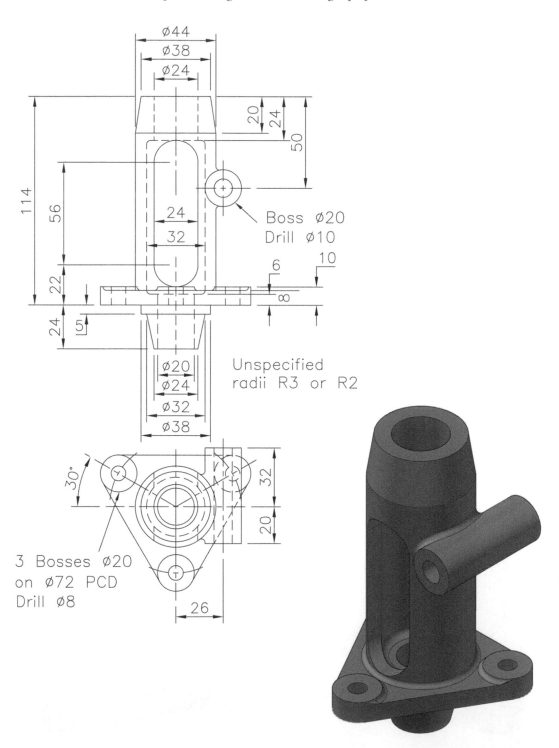

Ø44
Ø38
Ø24
20
24
50
114
56
24
32
Boss Ø20
Drill Ø10
6   10
22
8
24   5
Ø20
Ø24
Ø32
Ø38

Unspecified
radii R3 or R2

30°
32
20

3 Bosses Ø20
on Ø72 PCD
Drill Ø8

26

Exercise O16: tappet lever

Two First Angle projection elevations to create a model of the tappet lever. Discretion required with this model.

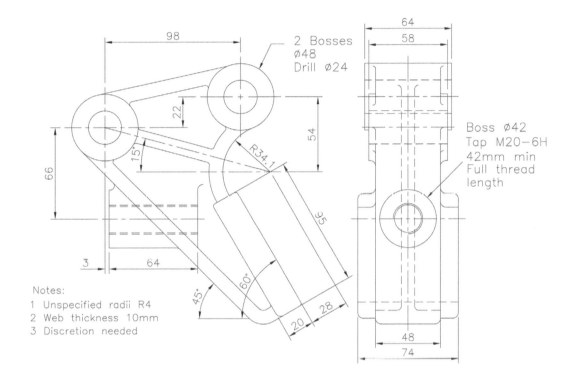

Notes:
1 Unspecified radii R4
2 Web thickness 10mm
3 Discretion needed

2 Bosses
Ø48
Drill Ø24

Boss Ø42
Tap M20-6H
42mm min
Full thread
length

Exercise O17: double mounting bracket

Two First Angle projection views to create the model.

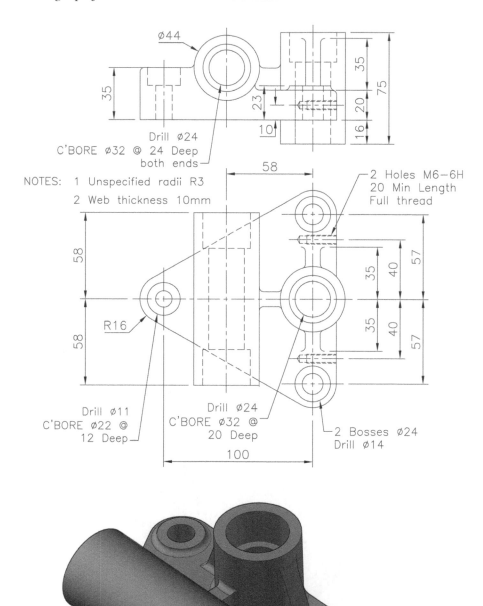

Ø44

35

35

75

23

20

10

16

Drill Ø24
C'BORE Ø32 @ 24 Deep
both ends

58

2 Holes M6–6H
20 Min Length
Full thread

NOTES: 1 Unspecified radii R3
2 Web thickness 10mm

58

58

R16

57

35

40

35

40

57

Drill Ø11
C'BORE Ø22 @
12 Deep

Drill Ø24
C'BORE Ø32 @
20 Deep

2 Bosses Ø24
Drill Ø14

100

Exercise O18: crane hook

1   Use the basic information from the sketch below to create a crane hook model.

2   Note that the 'tip end' of the hook has been left for you to complete.

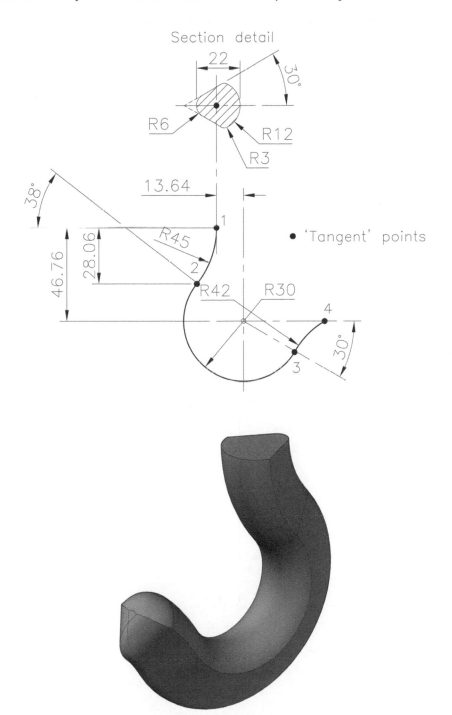

Exercise O19: lathe steady casting

1   Create a model of the lathe steady casting using the two Third Angle projection elevations displayed.

2   *Note*:
   (a)   Feature A: 2 Bosses Ø22, Tap M8-6H THRU to Ø16 hole.
   (b)   Feature B: 2 Holes M8-6H, 30 Min Length Thread, Drill Minor Dia THRU.
   (c)   Feature C: 2 Holes Ø16 THRU.
   (d)   FACES X and Y are similar.
   (e)   General 'Web' thickness is 6mm.
   (f)   Unspecified radii to be 3mm (or any other suitable value).
   (g)   Hidden detail has been omitted from faces X and Y.

Note: This is a relatively difficult model to complete, especially adding the several 6mm 'webs', so good luck with it.

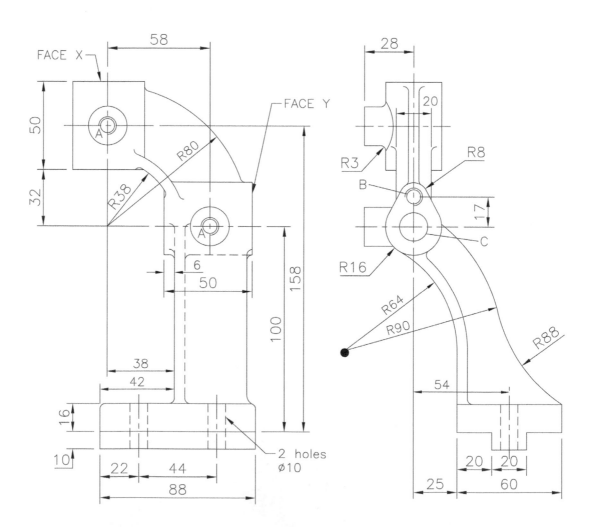

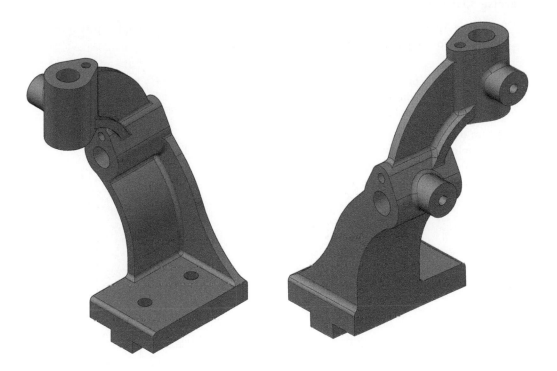

Exercise O20: gearbox cover

1   Two First Angle views of a gearbox cover to create a reasonably complex part model, with your discretion required for some dimensions and layout.

2   *Note*:
 (a)   all webs are 6mm;
 (b)   unspecified radii R3; and
 (c)   general wall thickness is 6mm.

3   Feature detail:
 (a)   6 holes M10-6H spaced as shown, with 3 on ⌀120 PCD.
 (b)   M20-5H, C'BORE ⌀25 × 16 deep.
 (c)   4 holes M8-6H, 10Min Length Full Thread.

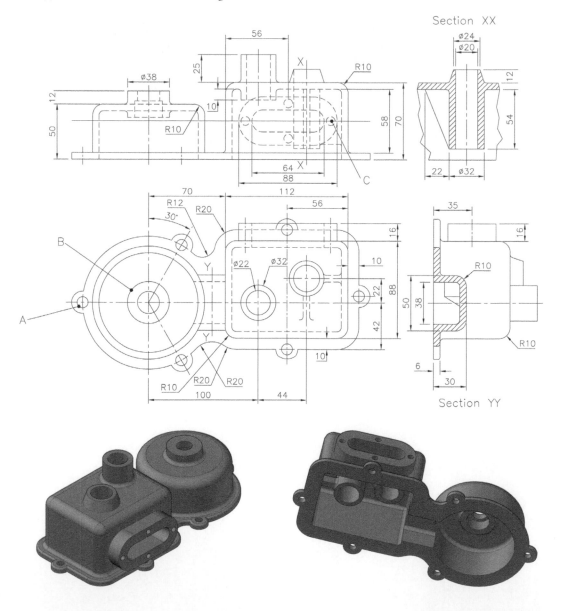

Exercise O21: valve body

1   A Third Angle projection layout for the model creation of the valve body, and note:
    (a) unless stated, the general radii to be 6mm; and
    (b) the single web thickness is 10mm.

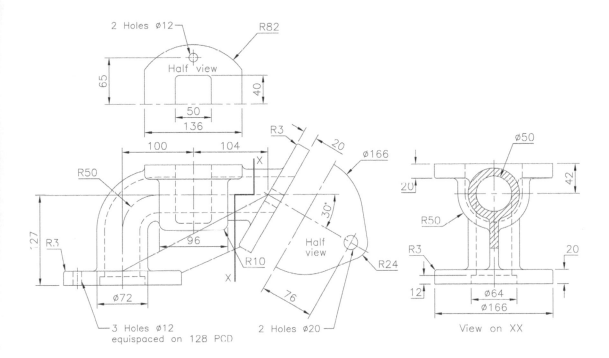

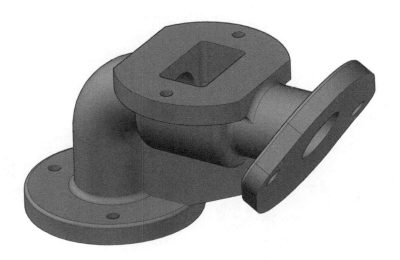

Exercise O22: machine vice base

Three First Angle views from which to create the component.

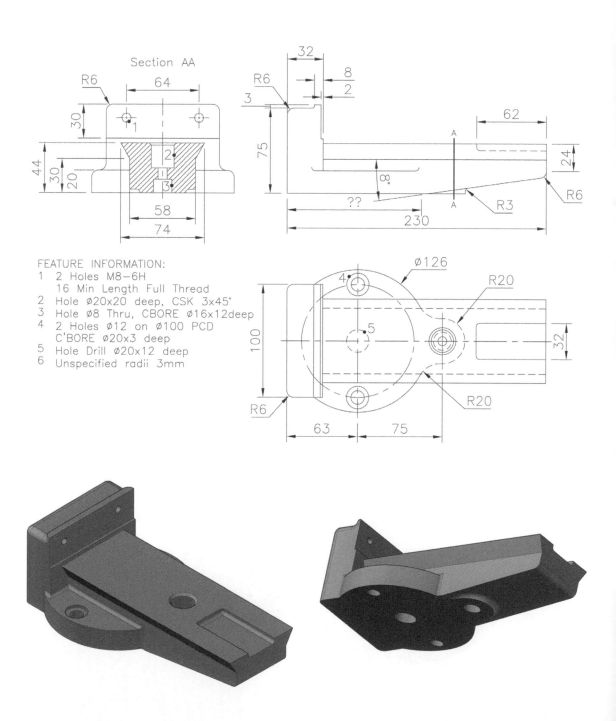

FEATURE INFORMATION:
1   2 Holes M8−6H
     16 Min Length Full Thread
2   Hole Ø20x20 deep, CSK 3x45°
3   Hole Ø8 Thru, CBORE Ø16x12deep
4   2 Holes Ø12 on Ø100 PCD
     C'BORE Ø20x3 deep
5   Hole Drill Ø20x12 deep
6   Unspecified radii 3mm

Exercise O23: cylinder

1   Two First Angle views (with additional views) for the creation of the component.
2   *Note*:
   (a)   the single web has a thickness of 12mm; and
   (b)   fillet radii to be R2 or R3 for smaller fillets and R6 for the elliptical flanges, or select
         your own fillet radii values.

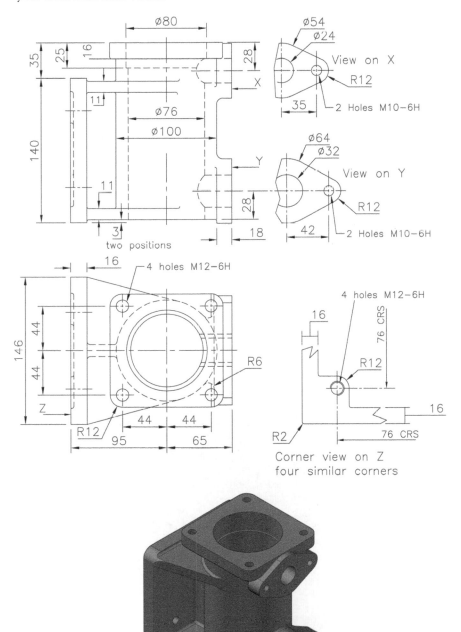

Exercise O24: junction box

A front elevation and two end elevations (one with a centre section view) in Third Angle projection are displayed for a model to be created with:

1   general wall thickness 6mm; and

2   unspecified radii 3mm, or use a suitable value.

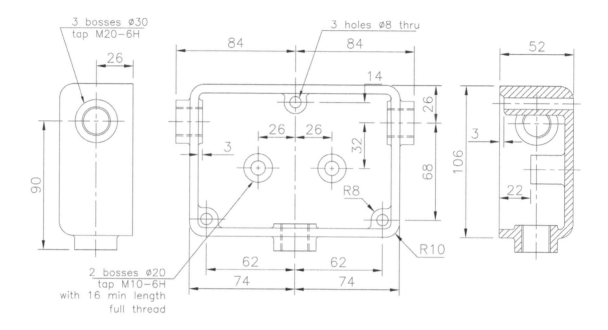

Exercise O25: bearing

Use the information from two Third Angle projection elevations (and a small auxiliary) to create the model with:

1   general wall thickness 9mm; and

2   unspecified radii 3mm, or use a suitable value.

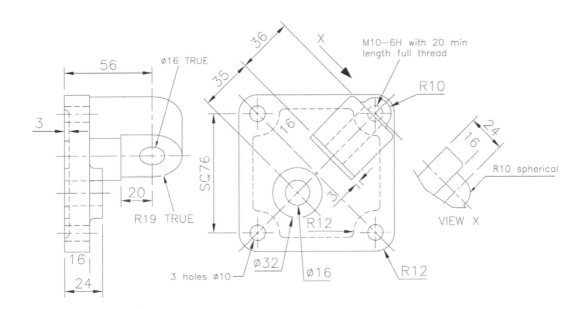

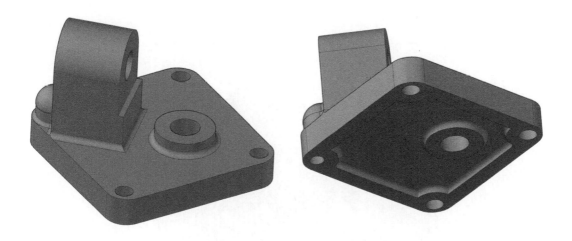

Exercise O26: anchor bracket

Use the two First Angle views to create a model of the component with unspecified radii 3mm or your own value.

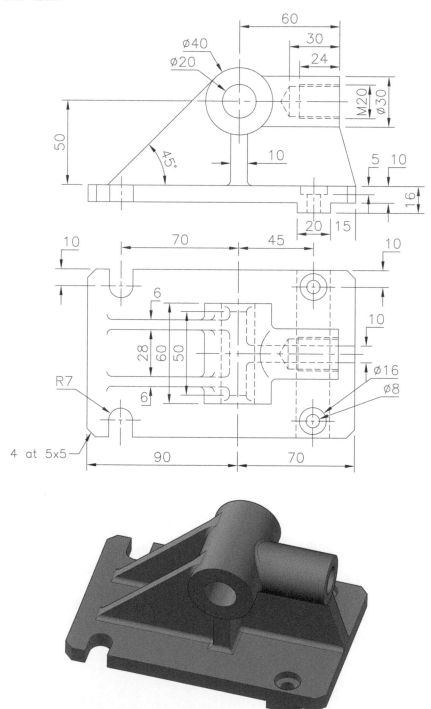

Exercise O27: bracket

Two First Angle views to create a model of the component.

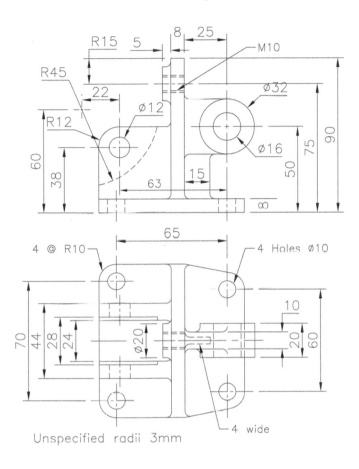

Unspecified radii 3mm

Exercise O28: hanging spindle

Create a model of the hanging spindle from the three Third Angle views displayed.

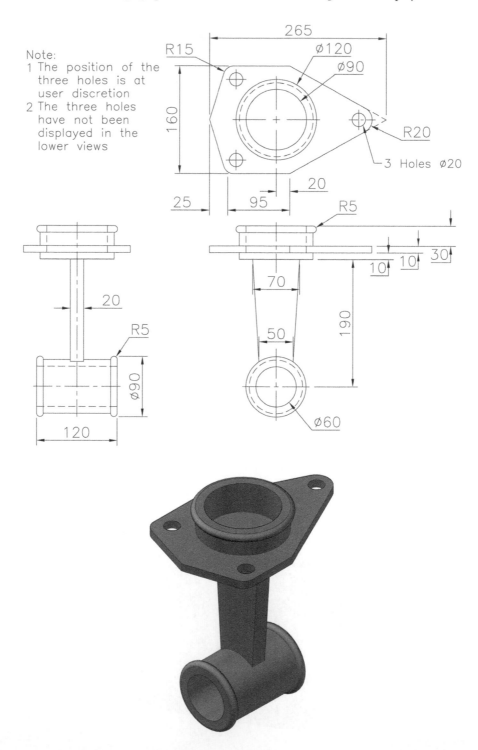

Note:
1 The position of the three holes is at user discretion
2 The three holes have not been displayed in the lower views

Exercise O29: bearing bracket

Two First Angle projection views for this model.

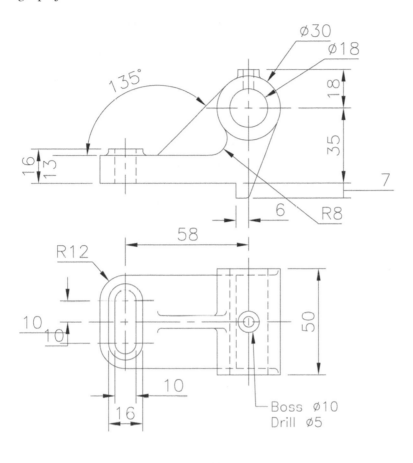

Exercise O30: bearing mounting

Another two First Angle projection views to create the model with unspecified radii 3mm.

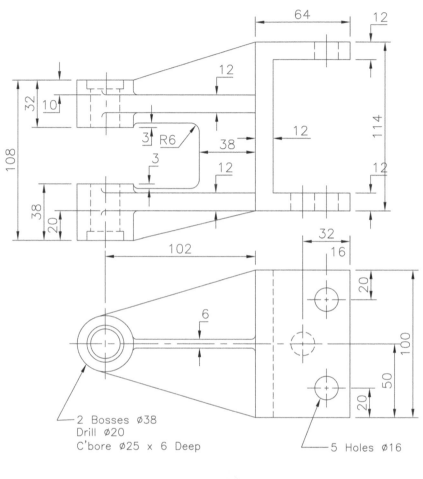

2 Bosses ⌀38
Drill ⌀20
C'bore ⌀25 x 6 Deep

5 Holes ⌀16

Exercise O31: column stand

A Third Angle elevation and half-section plan for the creation of the model.

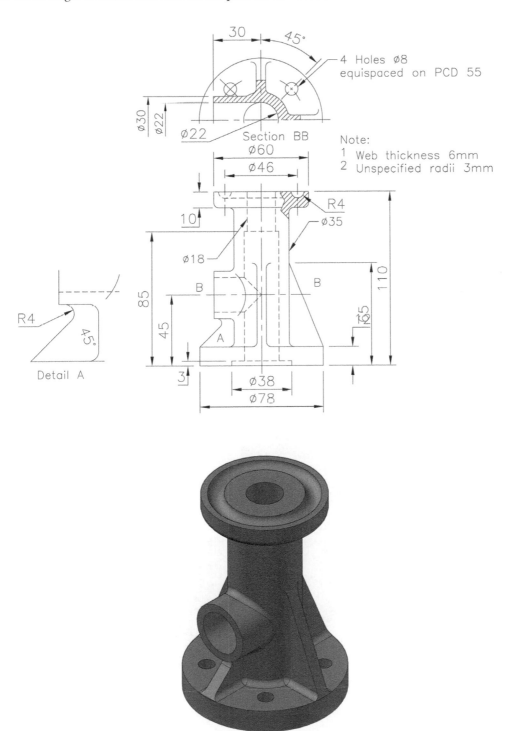

30 45°

4 Holes ⌀8
equispaced on PCD 55

⌀30 ⌀22

⌀22 Section BB

⌀60

⌀46

Note:
1 Web thickness 6mm
2 Unspecified radii 3mm

10

R4

⌀35

⌀18

85

B B

110

45

55

A

R4 45°

Detail A

3

⌀38

⌀78

Exercise O32: housing fixture bracket

Two Third Angle elevations for the creation of the model.

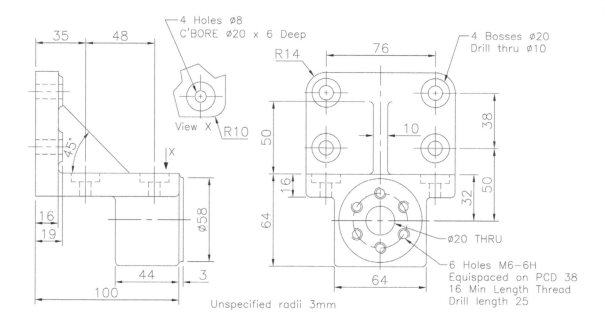

4 Holes ∅8
C'BORE ∅20 x 6 Deep

View X

R10

4 Bosses ∅20
Drill thru ∅10

∅20 THRU

6 Holes M6−6H
Equispaced on PCD 38
16 Min Length Thread
Drill length 25

Unspecified radii 3mm

Exercise O33: steam valve elbow

Two partially completed Third Angle views and two part 'auxiliaries' for the creation of the model.

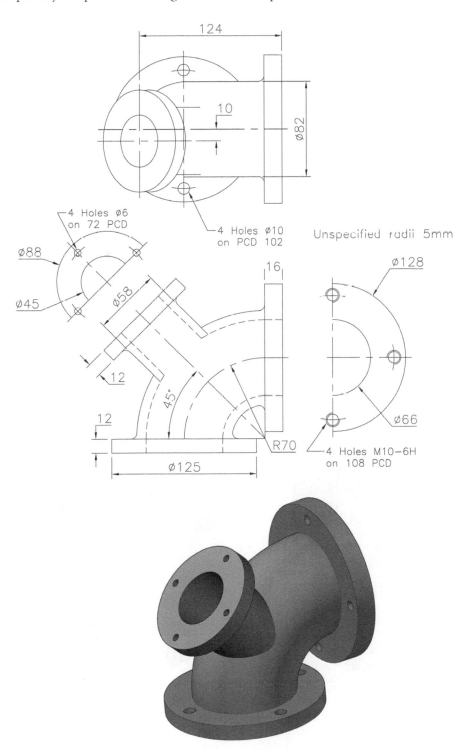

Exercise O34: separator

Third Angle layout to create the model.

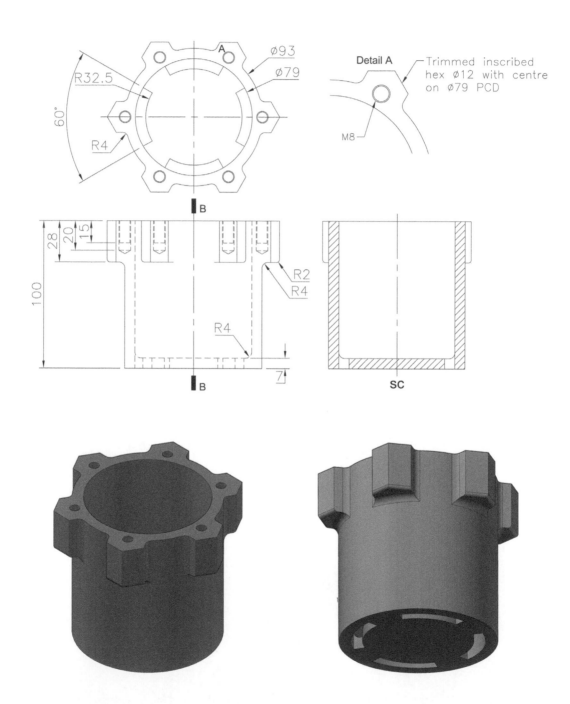

Exercise O35: sump

Two Third Angle views to create the model.

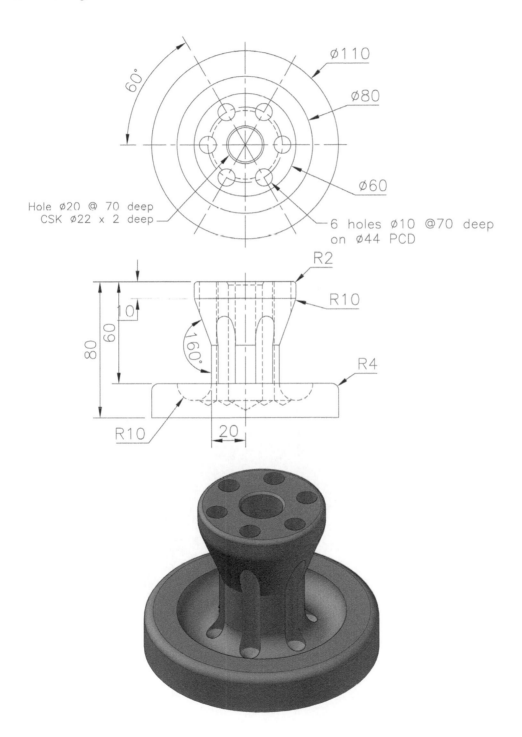

Exercise O36: divider

Traditional Third Angle layout for this model creation.

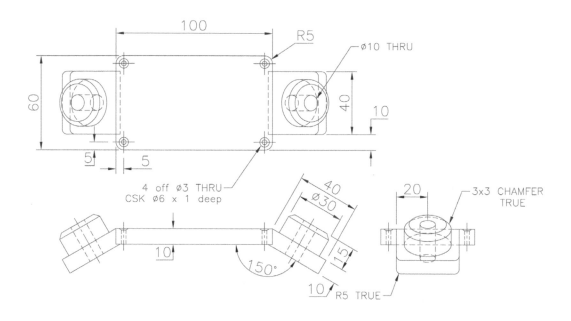

Exercise O37: spacer

Third Angle layout for this model creation, with your discretion required.

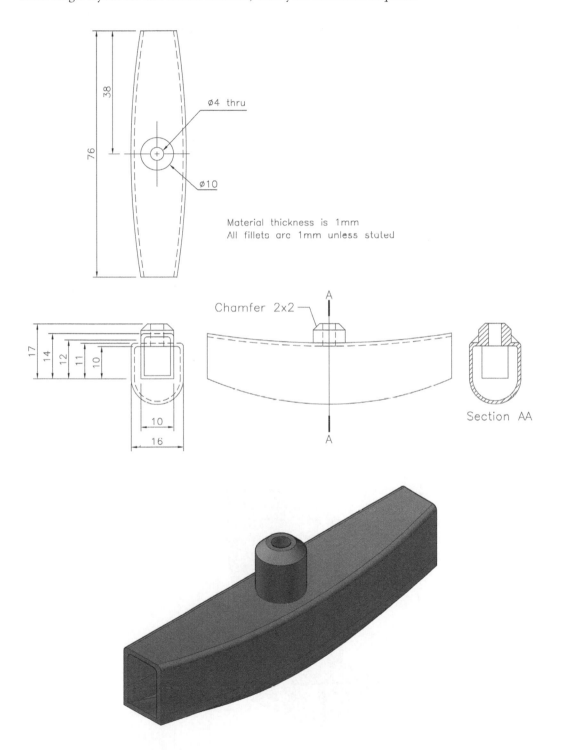

Ø4 thru

Ø10

Material thickness is 1mm
All fillets arc 1mm unless stated

Chamfer 2x2

Section AA

Exercise O38: turbine blade

1   The very basic outline of a turbine blade with twist and fir-tree root is displayed in the three views below.

2   Create a model of this single blade and save, as it may be used in a later exercise.

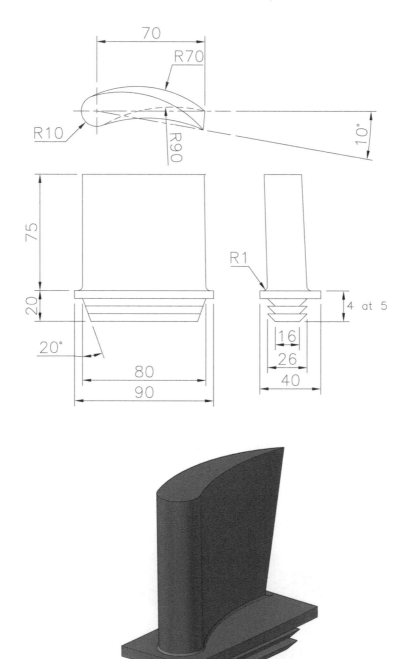

Exercise O39: compressor blade

1   The very basic outline of a compressor blade that has the same profile 'twisted twice' along
the blade length.

2   Create a model of this single blade and save, as it may be used in a later exercise.

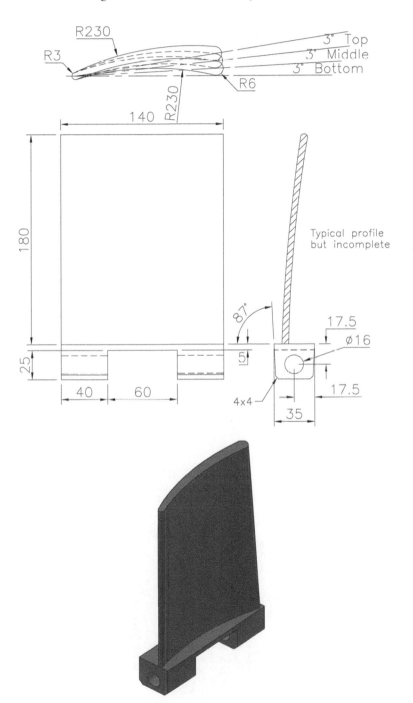

Exercise O40: rod end and eccentric shaft head

Create a model of the component using the two Third Angle drawings displayed.

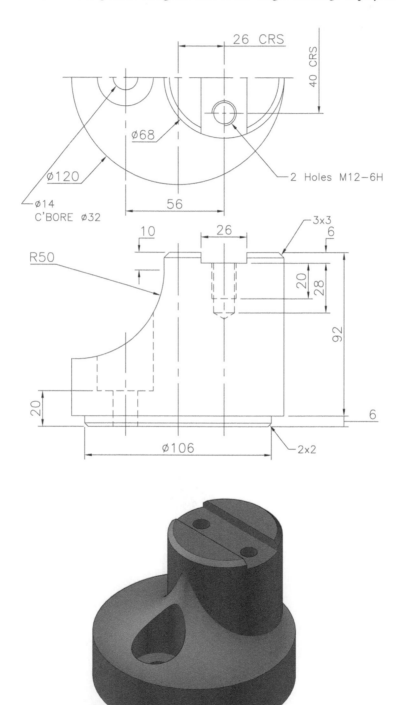

P   More complex models

1   In this chapter, I have included what I consider to be 'more complex models', although they may not be complex to the reader.

2   Also included are some assemblies that will be used for creating drawing layouts in Chapter 17.

3   The information will be given as First or Third Angle projection views, and as usual your discretion is required, especially for unspecified fillet radii.

4   Where possible, use the Library of standard parts.

5   *Note*:
   (a)   I have not displayed the actual models in this chapter, only the drawing views. However, all the completed models are displayed on the companion website (www.routledge. com/cw/mcfarlane), in the folder of sample files for Chapter 16.
   (b)   My idea is that the reader should attempt the creation of the various models from the orthographic views given, without the assistance of my model display.
   (c)   With assemblies, I usually display the parts with the listed material. If no suitable material is available, I display the parts with alternative materials or colours. This enhances the assembly appearance.
   (d)   I found that creating the models in this section was much easier and simpler than creating the 2D drawings.

Exercise P1: bearing assembly

1 An assembly exercise to start this section that consists of the following parts (part no: part name (no of, material)):
   (a) Part 1: Hanger Bracket (1, cast iron).
   (b) Part 2: Bearing Cap (1, cast iron).
   (c) Part 3: Top Bearing Step (1, brass).
   (d) Part 4: Bottom Bearing Step (1, brass).
   (e) Part 5: M20 Stud (2, nickel).
   (f) Part 6: M20 Hex Nut (2, as available from Library).
   (g) Part 7: M20 Washer (2, as available from Library).

2 Using the Third Angle drawing information provided, create a model for each part, then create the assembly.

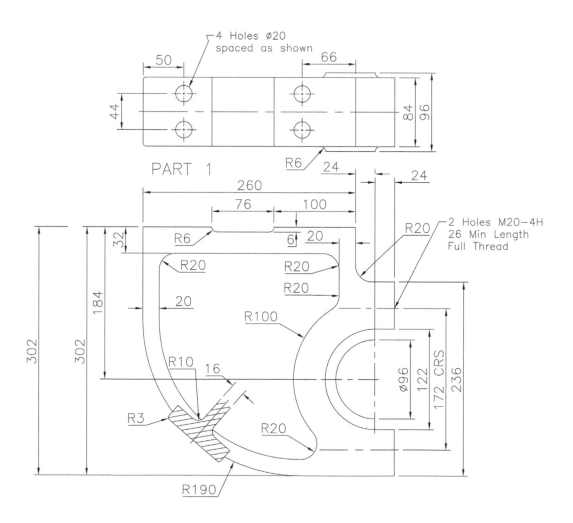

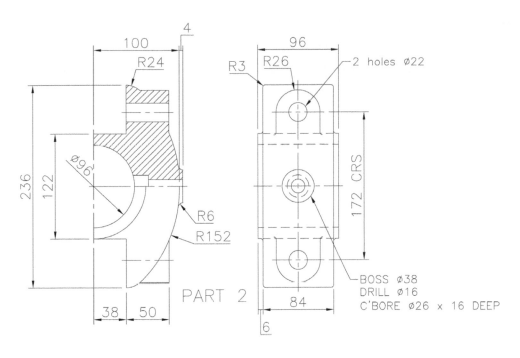

PART 2

BOSS ø38
DRILL ø16
C'BORE ø26 x 16 DEEP

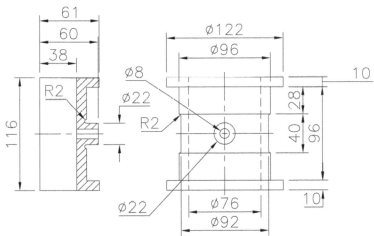

PART 3 : TOP
PART 4 : BOTTOM — identical to top
            but NO lubricator boss)

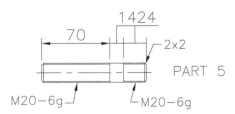

PART 5

M20—6g   M20—6g

Exercise P2: aircraft bracket

1   Two First Angle views of the component are displayed.

2   Note that the right view does not include all hidden detail, and my apologies for the actual dimensions displayed.

3   Other information:
   (a)   unspecified radii R2 or your own value;
   (b)   the webs are 4mm thick; and
   (c)   the three holes:
   •   X: M10-5H thru;
   •   Y: Ø6. C'BORE Ø12 × 3 deep; and
   •   Z: Ø6 thru.

4   Note that I have not included the fillet radii at every place in the 2D drawings.

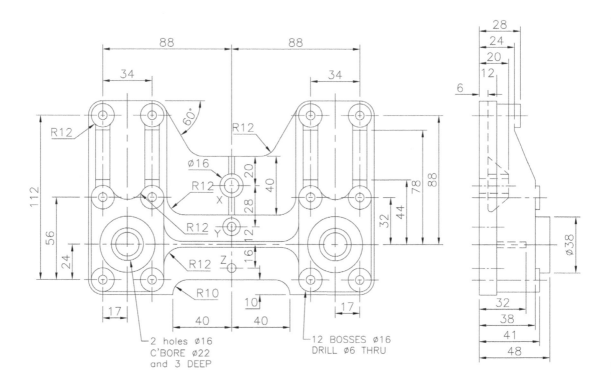

Exercise P3: oil pump cover

1   First Angle views of the component to create the model.

2   The five 'cut-outs' are quite involved, and you may require some patience to complete them.

3   The 'internal' radii of these cut-outs is 6mm.

4   The general wall and web thickness is 3mm and the general fillet radius is 1.5mm.

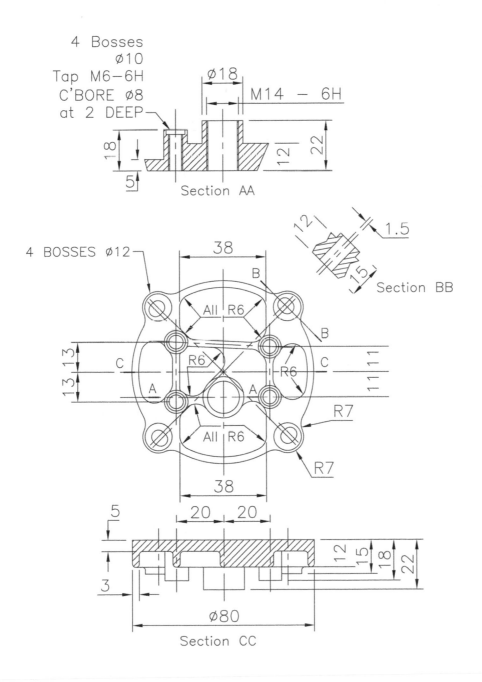

Exercise P4: carburettor body

1   Two First Angle projection views to create the model (which is a bit easier than the previous two models).

2   The following should be noted:
   (a)   the single web (in the 'middle') is 6mm wide; and
   (b)   the general fillet radii is 2.5mm (or your own suitable value), and may not be completely displayed in the views.

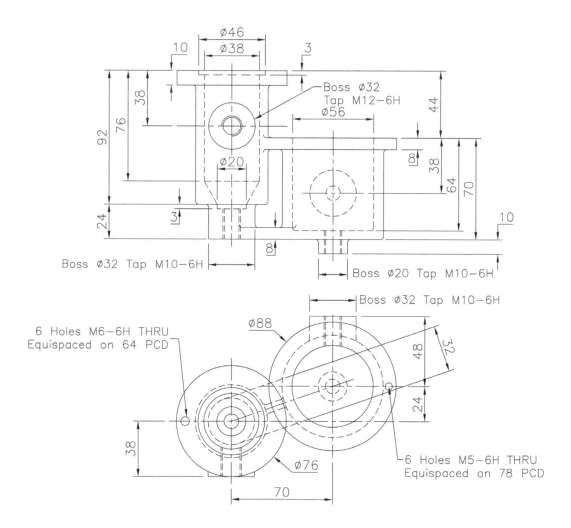

Exercise P5: clamp assembly

1 Several parts (in First Angle) to create for this assembly:

(a) Part 1: Fixed Jaw.
(b) Part 2: Moving Jaw.
(c) Part 3: Fixed Jaw Machined Plate.
(d) Part 4: Moving Jaw Machined Plate.
(e) Part 5: Steady
(f) Part 6: Adjusting Screw.

(g) Part 7: Tommy Bar.
(h) Part 8: 6 of M5-6g slotted CSK head screws 12 lg (or suitable).
(i) Part 9: 1 of M10-6g slotted cheese head screw 35 lg (or suitable).

2 Create each part using the drawing information as given, then create an assembly with parts 3 and 4 20mm apart and the 'Tommy Bar' horizontal.

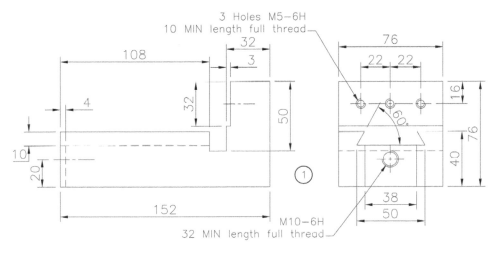

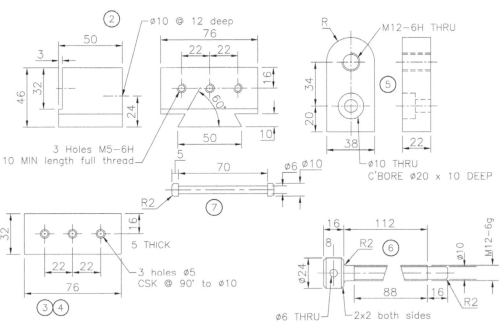

Exercise P6: engine mounting fitting

A single orthographic view to create the model.

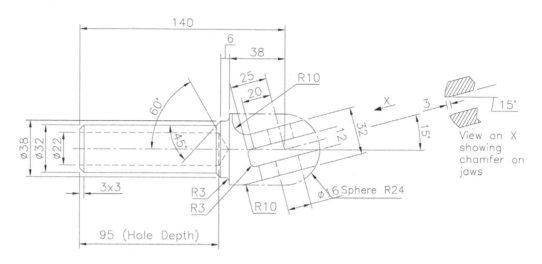

Exercise P7: location fixture

A plan and elevation in First Angle projection, as well as two 'half' end elevations to give details 'about each end' of the component.

The general wall/base and web thickness is 6mm.

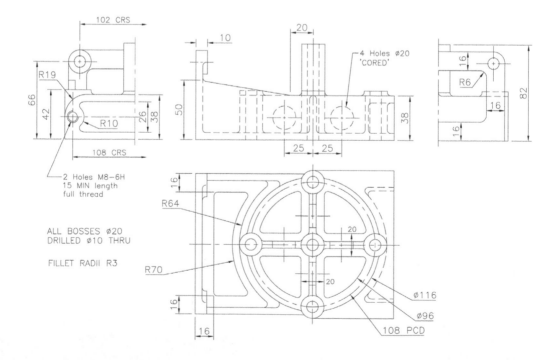

Exercise P8: pump barrel

1 A First Angle projection sectional elevation and an end elevation with other views to create the model.

2 No 'real' problems with this model, but be careful 'adding the various holes'.

3 The following notes are relevant:
(a) ★ 4 holes Ø14 equispaced on 102 PCD;
(b) flanges X and Y have the same outside diameter and hole pattern;
(c) fillet radii 6mm (where appropriate); and
(d) use your discretion for unsure sizes/position.

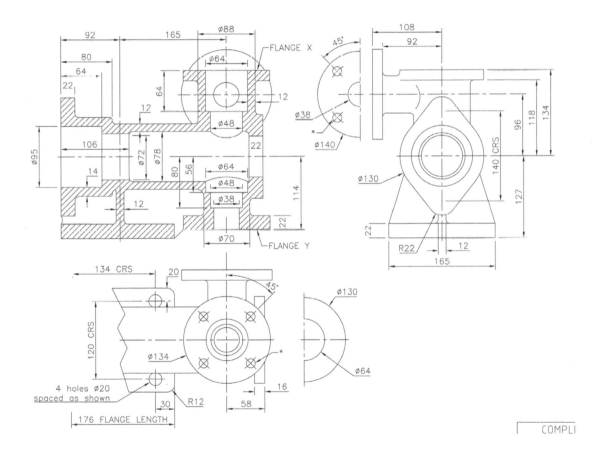

Exercise P9: distribution casing

1   Two Third Angle projection elevations to create the model.

2   Note the following:
   (a) fillet radii R3;
   (b) all walls and webs are 6mm thick; and
   (c) all holes are ⌀10mm unless otherwise specified.

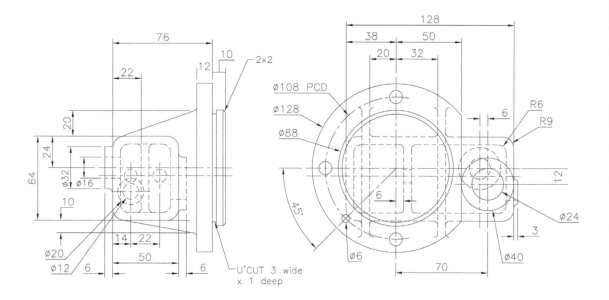

Exercise P10: belt adjuster assembly

1 A simple assembly consisting of:
   (a) Part 1: Base Plate.
   (b) Part 2: Adjusting Bracket.
   (c) Part 3: Adjusting Screw.
   (d) Part 4: Special Bolt.
   (e) Part 5: Set Screw.
   (f) Others: M12-6H Hex Nut and M12 Plain Washer.

2 *Note*: Edges XX and YY have to be in line in the assembly.

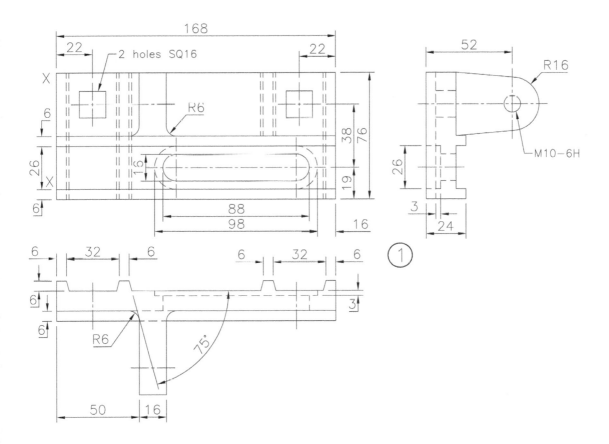

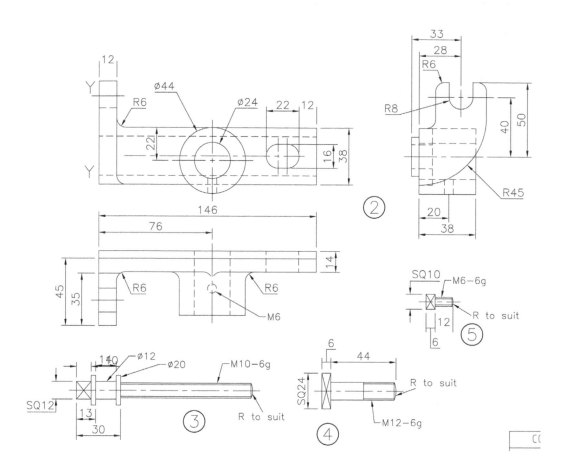

Exercise P11: valve casing

1   Two 'busy' First Angle views of the casing to be created.

2   *Note*:
  (a) Unspecified fillet radii 3mm (or other suitable).
  (b) The three equispaced webs have a thickness of 6mm.
  (c) The 'triangular' webs have a thickness of 10mm.
  (d) My apologies for the drawing 'clutter'. This was a very difficult drawing to add 'readable' dimensions, but hopefully you will be able to get the required sizes to complete the model and use your discretion as usual.
  (e) Not all small fillets are displayed, and the flange at 30 degrees is not my best.

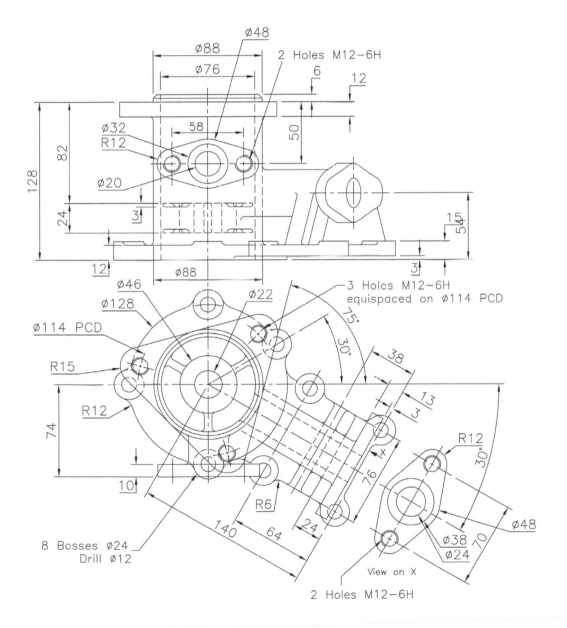

Exercise P12: drill jig body

A single First Angle plan view with two part section views for the creation of this model.

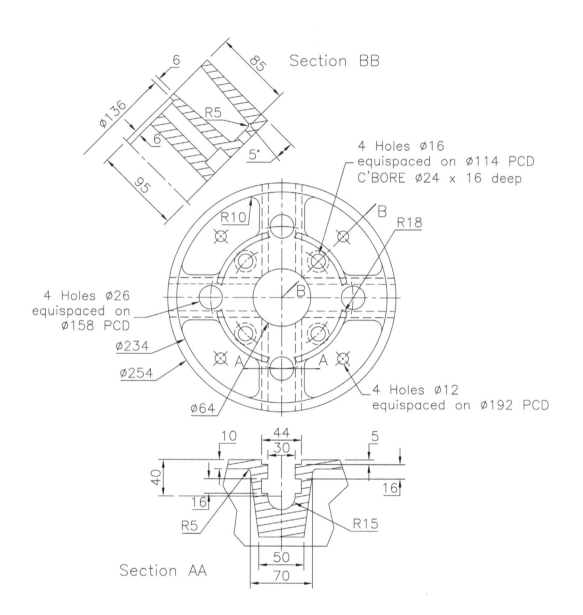

Exercise P13: connecting rod end assembly

An assembly required for the following parts:

1   Part 1: Steel Connecting Rod.

2   Part 2: Brass Bearing Halves (2 of).

3   Part 3: Mild Steel Front Plate.

4   Part 4: Mild Steel Locking Plate.

5   Part 5: Mild Steel Bolts (2 of).

6   Others:
    (a) two steel M22-6H Hex Nuts – Library; and
    (b) one M10-6g Hex Head Screw (26 lg) to secure the locking plate to the front plate.

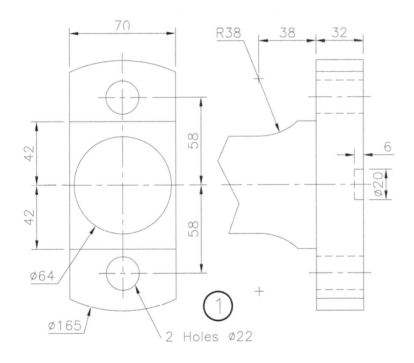

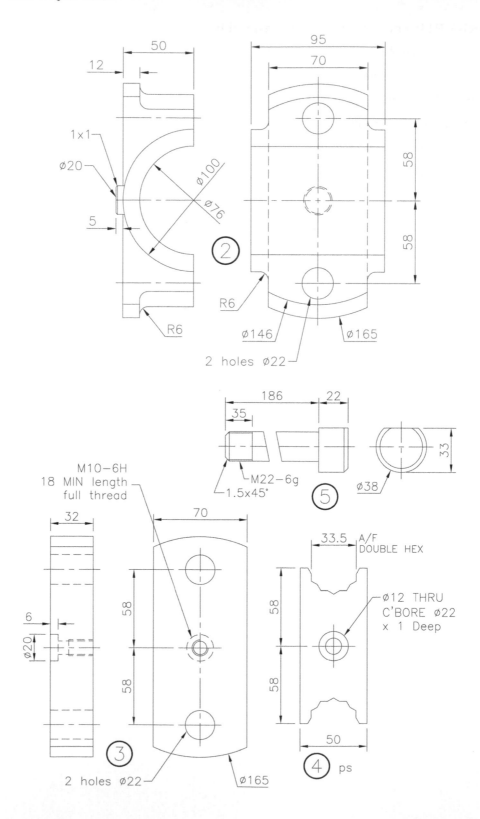

50

12

1×1

∅20

5

∅100

∅76

②

R6

R6

∅146

∅165

2 holes ∅22

95

70

58

58

186

22

35

M22−6g

1.5x45°

⑤

∅38

33

M10−6H
18 MIN length
full thread

32

70

33.5 A/F
DOUBLE HEX

6

∅20

58

58

③

2 holes ∅22

∅165

58

58

∅12 THRU
C'BORE ∅22
x 1 Deep

50

④ ps

Exercise P14: footstep bearing assembly

This assembly consists of the following parts, displayed in Third Angle projection.

1   Part 1: Cast Iron Base Plate.

2   Part 2: Cast Iron Bearing Support.

3   Part 3: Bronze Bearing Bush.

4   Part 4: Bronze Bearing Pad.

5   Part 5: Square Head Bolts (4 of).

6   Part 6: Steel Dowel (1 of).

7   Others:
    (a)   4 of M8-6H Hex Nuts − Library; and
    (b)   4 of M8 plain washers − Library.

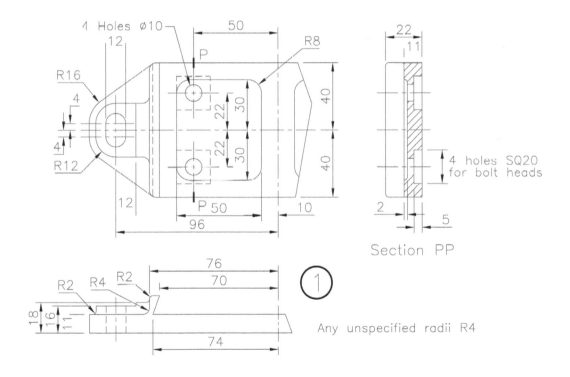

Section PP

Any unspecified radii R4

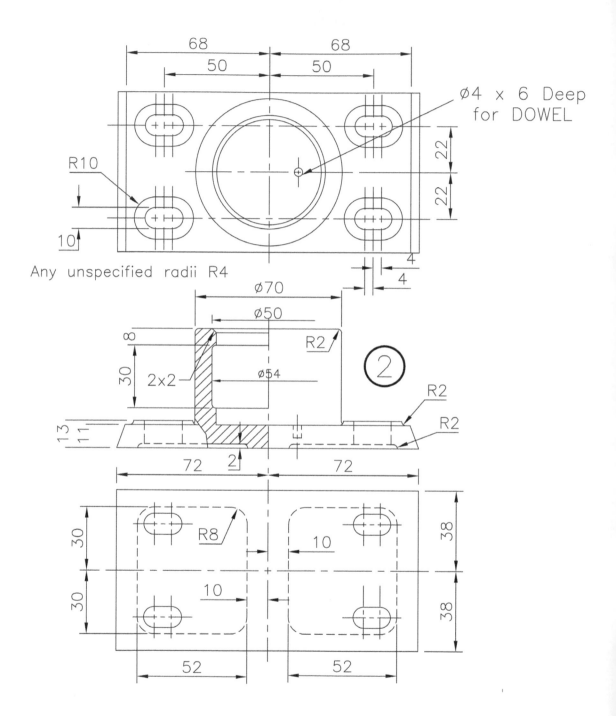

Ø4 x 6 Deep
for DOWEL

R10

Any unspecified radii R4

Ø70

Ø50

R2

2x2

Ø54

R2

R2

13
11

72   2   72

R8

10

30

10

30

38

38

52   52

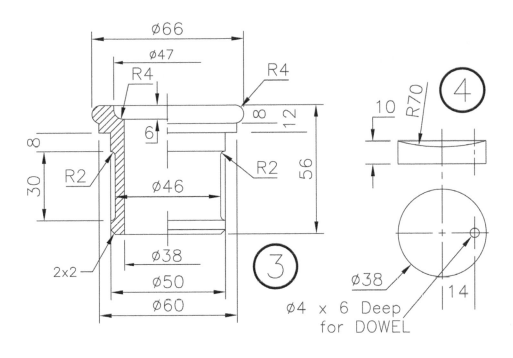

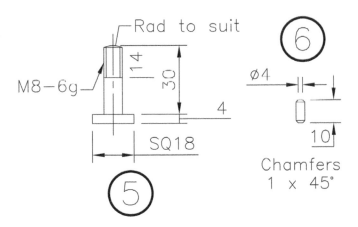

Ø66

Ø47

R4

R4

8

12

6

56

8

30

R2

R2

Ø46

Ø38

2x2

Ø50

Ø60

③

④

10

R70

Ø38

14

Ø4 x 6 Deep
for DOWEL

Rad to suit

⑥

14

30

M8−6g

4

Ø4

SQ18

10

⑤

Chamfers
1 x 45°

Exercise P15: control bracket

1   A slightly different type of drawing display for this exercise.

2   Two First Angle projection elevations with quite a lot of detail.

3   Also displayed are plan views of the three web extensions, the dimensions being relative to the main 'column' vertical centre line.

4   Hopefully, I have managed to include all dimensions, but your discretion is advised.

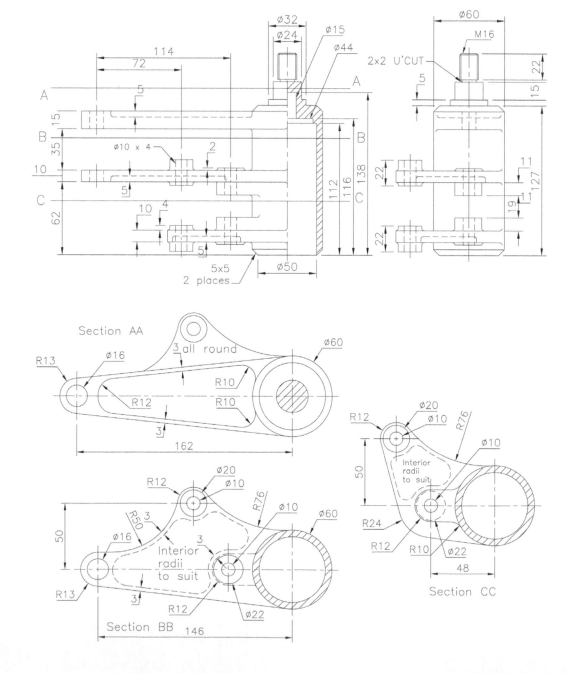

Exercise P16: scoop

Using the First Angle elevation and plan views below, create a model of the scoop, which is made from galvanised material with a thickness of 2mm.

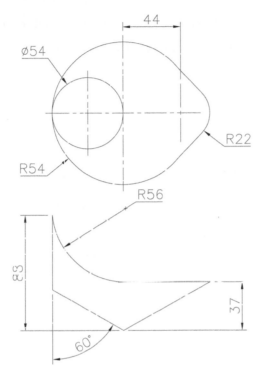

Exercise P17: oblique plane block

1 The plan and elevation (First Angle) of a composite block is displayed below.

2 The block is 'cut' by an oblique plane, the vertical and horizontal traces being shown by VT and HT.

3 Create a model of the cut block.

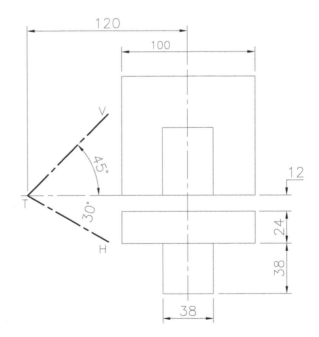

Exercise P18: lathe tool holder

Create an assembly of a lathe tool holder consisting of the following parts:

1   Part 1: steady (1 of);

2   Part 2: holder (1 of, partial dimensions given – design the ends to suit);

3   Part 3: tightening screw (1 of, partial dimensions given – design to suit); and

4   Part 4: tool tip (2 of, design to suit).

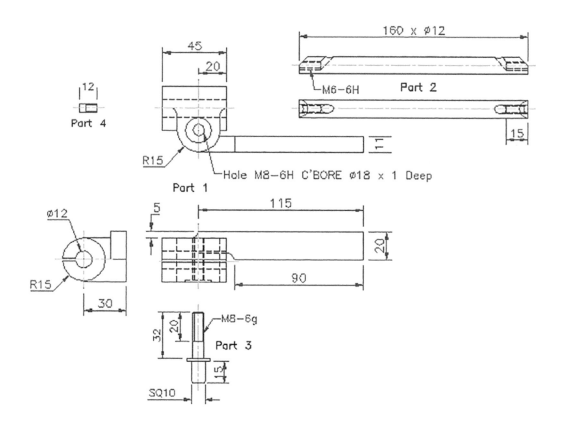

Exercise P19: valve mechanism

1 An assembly is required using the drawing details displayed.

2 The parts in the assembly are:
   (a) Part 1: Shaft (1);
   (b) Part 2: Link (1);
   (c) Part 3: Rod (1);
   (d) Part 4: Shaft Bush (2);
   (e) Part 5: Rod Bush (1);
   (f) Part 6: Pin (1); and
   (g) Part 7: Split Pin (1).

3 Create the assembly using the various parts.

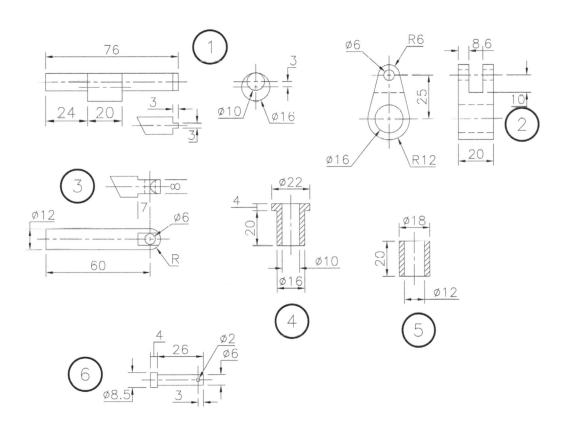

Exercise P20: handrail clamp assembly

1   A small handrail clamp consists of a top and bottom part with two bolts of appropriate length.

2   Using the drawing information below, create each part and then create an assembly of the handrail clamp.

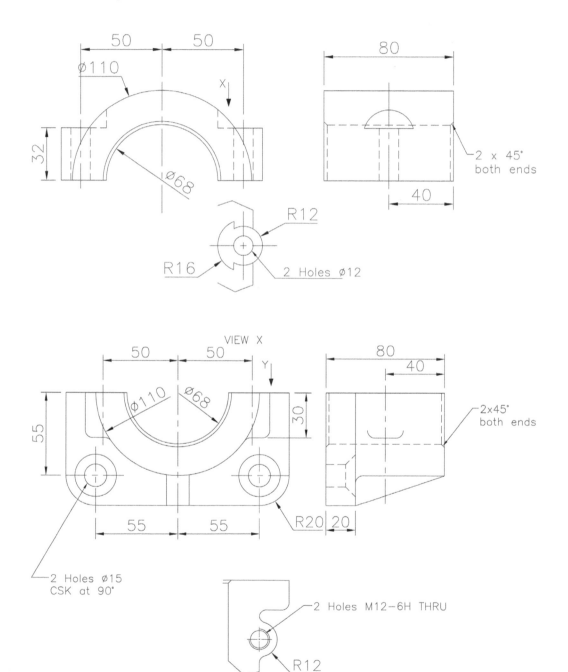

Exercise P21: wall–mounted hook

1   A crane hook is attached to a wall-mounted elliptical flange.

2   Create an assembly consisting of the following single parts:
   (a) elliptical wall-mounted plate;
   (b) hook;
   (c) M12 Hex Headed Bolt of suitable length; and
   (d) M12 Hex Nut.

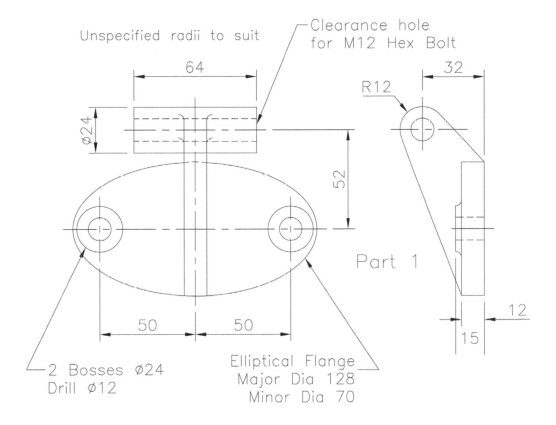

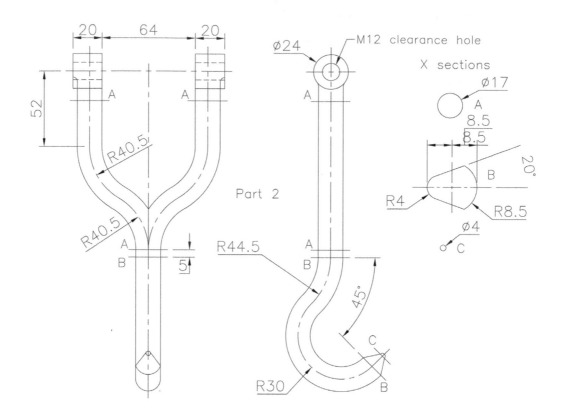

Part 2

M12 clearance hole

X sections

Ø24

Ø17

R44.5

R30

R40.5

R40.5

52

20   64   20

5

R4

R8.5

Ø4

20°

8.5

8.5

45°

Exercise P22: cover

1   Two Third Angle projection views for the creation of this model.

2   *Note*:

   (a)   wall thickness is 6mm, unless otherwise stated;

   (b)   outside fillet radii is 5mm, unless otherwise stated; and

   (c)   inside fillet radii is 3mm.

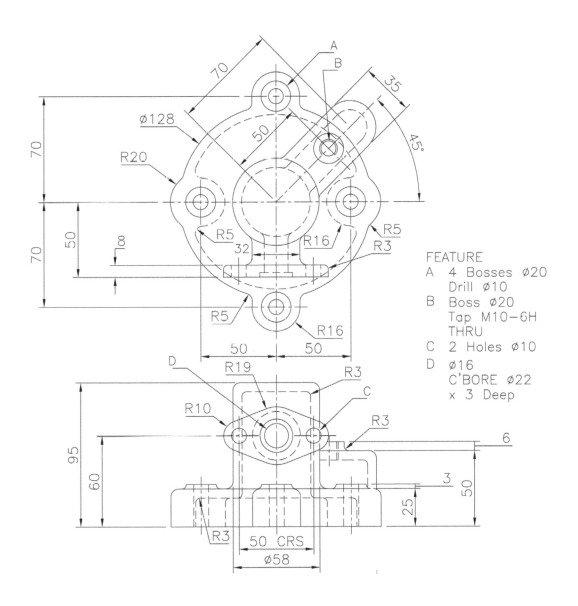

FEATURE

A   4 Bosses ⌀20
    Drill ⌀10

B   Boss ⌀20
    Tap M10−6H
    THRU

C   2 Holes ⌀10

D   ⌀16
    C'BORE ⌀22
    x 3 Deep

Exercise P23: bevel gear

1   The cross-section of a bevel gear 'disc' is detailed below.

2   Also displayed is the bevel gear tooth profile.

3   Create a model of the bevel gear with your own design for the number of teeth.

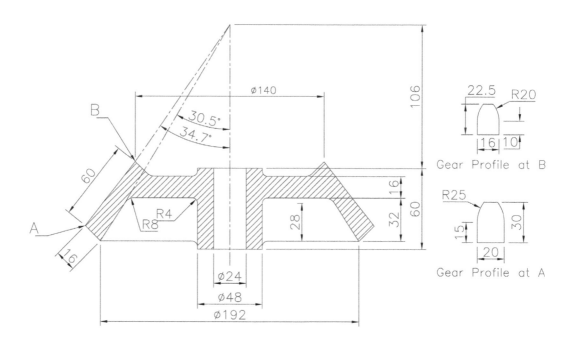

Gear Profile at B

Gear Profile at A

Exercise P24: wall bracket

1   Three First Angle views of the component to be created are displayed.

2   The two elevations do not give the 'complete picture', but there are two auxiliary views to assist.

3   Of all the exercises in the book, this was the one that gave me the most problems, as it took me some time to figure out the 2D drawing views (which I have reproduced exactly), and the 'sloped view B' part took me some time to reason out and complete.

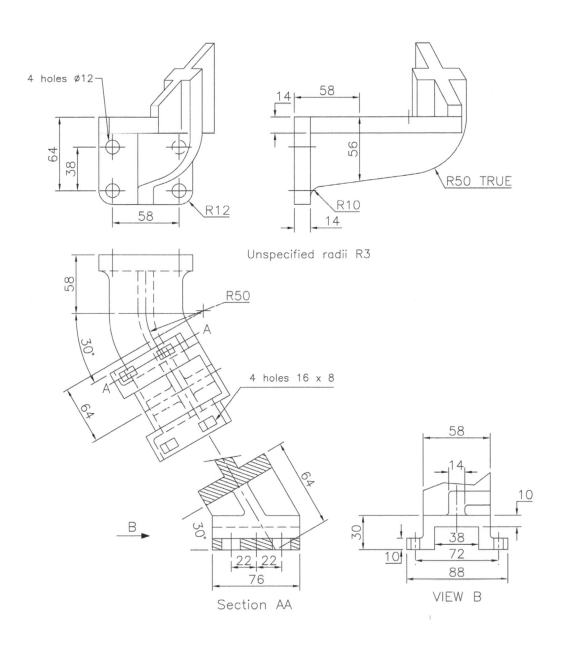

Q Detail drawings

1   The name of this chapter is a bit misleading as there are no additional models to create.

2   Basically, it is producing drawing layouts of models previously created, so it is your decision whether to proceed with the chapter or not.

3   At this stage, the reader should have the knowledge and ability to create drawing layouts in both First and Third Angle projection, so no instructions should be needed, other than that assemblies should include a parts list.

4   Only four drawing layouts are displayed, these being:
(a)   O15: indicator sleeve;
(b)   O20: gearbox cover;
(c)   P14: footstep bearing assembly; and
(d)   P21: wall-mounted hook assembly.

5   The two assemblies display both a detailed (with parts list) and exploded layout.

6   Suggested other models for creating drawing layouts are displayed on the companion website (www.routledge.com/cw/mcfarlane), in the folder of sample files for Chapter 17, as follows:

(a)   *Single models*
- O6: bearing
- O8: base block
- O9: crank disc
- O10: sealing cap
- O11: clutch bracket
- O12: spindle housing
- O13: intermediate casing
- O16: tappet lever
- O17: double mounting bracket
- O19: lathe steady casting
- O21: valve body
- O23: cylinder
- O28: hanging spindle
- O30: bearing mounting
- O31: column stand
- O34: separator
- P2: aircraft bracket
- P3: oil pump cover

- P4: carburettor body
- P6: engine mounting fitting
- P7: location fixture
- P8: pump barrel
- P9: distribution casing
- P11: valve casing
- P12: drill jig body
- P15: control bracket
- P22: cover
- P24: wall bracket

(b)   *Assemblies*
- P1: bearing assembly
- P5: clamp assembly
- P10: belt adjuster assembly
- P13: connecting rod end assembly
- P18: lathe tool holder
- P19: valve mechanism
- P20: handrail clamp assembly

Exercise Q1: indicator sleeve (O15)

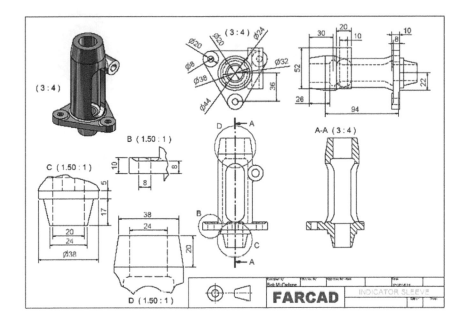

Exercise Q2: gearbox cover (O20)

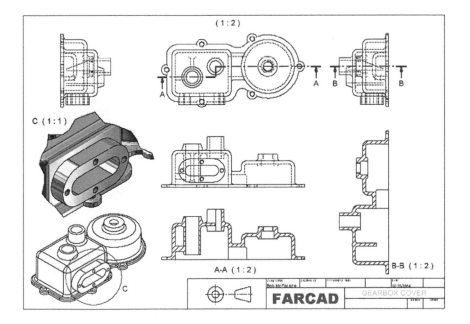

Exercise Q3: footstep bearing assembly (P14)

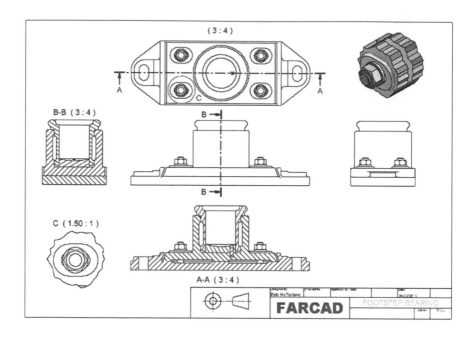

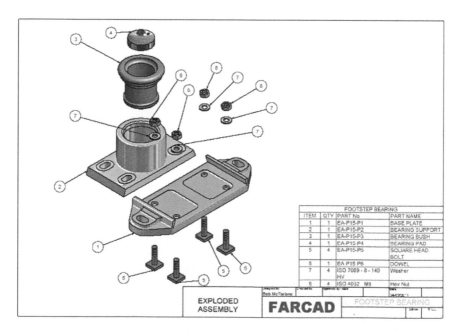

ITEM	QTY	PART No	PART NAME
		FOOTSTEP BEARING	
1	1	EA-P15-P1	BASE PLATE
2	1	EA-P15-P2	BEARING SUPPORT
3	1	EA-P15-P3	BEARING BUSH
4	1	EA-P15-P4	BEARING PAD
5	4	EA-P15-P5	SQUARE HEAD BOLT
6	1	EA P15 P6	DOWEL
7	4	ISO 7089 - 8 - 140 HV	Washer
8	4	ISO 4032 M8	Hex Nut

EXPLODED ASSEMBLY

FARCAD

FOOTSTEP BEARING

Exercise Q4: wall–mounted hook (P21)

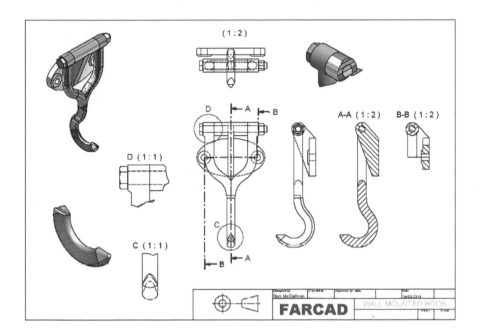

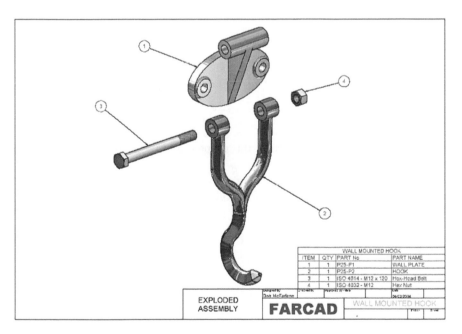

ITEM	QTY	PART No	PART NAME
1	1	P25-P1	WALL PLATE
2	1	P25-P2	HOOK
3	1	ISO 4014 - M12 x 120	Hex-Head Bolt
4	1	ISO 4032 - M12	Hex Nut

EXPLODED ASSEMBLY

FARCAD

WALL MOUNTED HOOK

R Material and physical properties

1 Inventor® allows both physical and material properties of part models and assemblies to be displayed in and/or extracted into other software packages.

2 The physical properties that can be obtained include:
   (a) centre (center) of gravity;
   (b) mass;
   (c) volume;
   (d) surface area;
   (e) mass moment of inertia; and
   (f) principle moment of inertia.

3 The material properties include:
   (a) density;
   (b) Young's modulus;
   (c) Poisson's ratio;
   (d) yield strength;
   (e) ultimate tensile strength;
   (f) thermal conductivity;
   (g) linear expansion; and
   (h) specific heat.

4 There are 12 exercises (all P models) in this chapter, and for each exercise you should:
   (a) open the named part model;
   (b) obtain the properties identified;
   (c) compare your values with those listed; and
   (d) save.

Exercise R1: aircraft bracket (P2)

For the first exercise in this chapter, I have:

1 assigned the material aluminium 6061 to the component;

2 displayed the centre of gravity icon and its associated X, Y and Z values in pictorial form; and

3 listed the other properties.

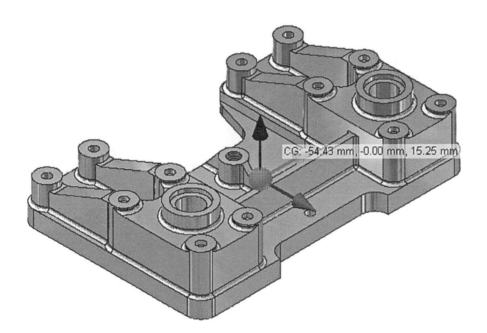

Material assigned: Aluminium 6061
Density: 2.710E-006 kg/mm^3
Centre of Gravity (all mm):
X: −54.414
Y: −5.068E-005
Z: 15.251
Mass: 1.240 kg
Volume: 8.990E+004 mm^3
Ixx: 5.703E+003 kg mm^2
Iyy: 1.630E+003 kg mm^2
Izz: 7.001E+003 kg mm^2

Exercise R2: oil pump cover (P3)

Material assigned: Mild steel
Density: 7.860E-006 kg/mm^3
Centre of Gravity (all mm):
X: 0.411
Y: −0.365
Z: 5.133E+004
Mass: 0.403 kg
Volume: 5.133E+004 mm^3
Ixx: 178.266 kg mm^2
Iyy: 188.827 kg mm^2
Izz: 353.197 kg mm^2

Exercise R3: carburettor body (P4)

Material assigned: Mild steel
Density: 7.860E-006
Centre of Gravity:
X: 36.666
Y: 14.001
Z: 48.479
Mass: 2.204
Volume: 2.804E+005
Ixx: 3.868E+003
Iyy: 5.921E+003
Izz: 4.647E+003

Exercise R4: engine mounting fitting (P6)

Material assigned: Mild steel
Density: 7.860E-006
Centre of Gravity:
X: −1.609E-006
Y: 92.740
Z: −1.898
Mass: 0.816
Volume: 1.038E+005
Ixx: 1.503E+003
Iyy: 1.006E-004
Izz: 2.869E-005

Exercise R5: location fixture (P7)

Material assigned: Aluminium 6061
Density: 2.710E-006
Centre of Gravity:
X: −61.066
Y: 90.699
Z: 20.155
Mass: 0.906
Volume: 3.343E+005
Ixx: 2.833E+003
Iyy: 1.891E+003
Izz: 4.171E+003

Exercise R6: pump barrel (P8)

Material assigned: Mild steel
Density: 7.860E-006
Centre of Gravity:
X: −7.869
Y: 170.864
Z: −20.038
Mass: 26.322
Volume: 3.350E+006
Ixx: 4.384E+005
Iyy: 2.168E+005
Izz: 3.227E+005

Exercise R7: distribution casing (P9)

Material assigned: Mild steel
Density: 7.860E-006
Centre of Gravity:
X: 10.831
Y: 13.407
Z: −3.582
Mass: 3.287
Volume: 4.182E+005
Ixx: 4.663E+003
Iyy: 6.590E+003
Izz: 6.693E+003

Exercise R8: valve casing (P11)

Material assigned: Mild steel
Density: 7.860E-006
Centre of Gravity:
X: 19.474
Y: 27.948
Z: 48.454
Mass: 4.645
Volume: 5.910E+005
Ixx: 1.938E+003
Iyy: 1.440E+004
Izz: 1.901E+004

Exercise R9: drill jig body (P12)

Material assigned: Aluminium 6061
Density: 2.710E-006
Centre of Gravity:
X: −0.001
Y: 0.696
Z: −47.985
Mass: 6.432
Volume: 2.374E+006
Ixx: 3.384E+004
Iyy: 3.349E+004
Izz: 5.781E+004

Exercise R10: control bracket (P15)

Material assigned: Mild steel
Density: 7.860E-006
Centre of Gravity:
X: −3.645
Y: −40.047
Z: 76.605
Mass: 2.377
Volume: 3.024E+005
Ixx: 9.256E+003
Iyy: 4.600E+003
Izz: 6.462E+003

Exercise R11: cover (P22)

Material assigned: Mild steel
Density: 7.860E-006
Centre of Gravity:
X: 5.416
Y: 2.765
Z: 32.974
Mass: 2.339
Volume: 2.975E+006
Ixx: 5.097E+003
Iyy: 4.249E+003
Izz: 6.276E+003

Exercise R12: wall bracket (P24)

Material assigned: Mild steel
Density: 7.860E-006
Centre of Gravity:
X: 73.317
Y: 14.935
Z: −3.691
Mass: 3.163
Volume: 4.024E+005
Ixx: 5.171E+003
Iyy: 1.033E+004
Izz: 8.432E+003

S Additional assemblies

1   For this chapter, there are 25 new assemblies to be created, and the procedure should be quite familiar to the reader:
    (a) Create each part as a new part model using the standard metric (mm) .ipt file.
    (b) Save each part when completed (adding material if required).
    (c) Create an assembly using the standard metric (mm) .iam file.
    (d) Use the Library for nuts, washers, bolts, etc.
    (e) Create a drawing layout of the assembly with the views to suit your own requirements, but with a parts list included.
    (f) Create a presentation/animation to your own requirements.

2   The parts for each assembly will be listed and the appropriate drawing information displayed, but remember to use your discretion.

3   (a) Applying unspecified fillet radii can be quite boring and time-consuming, but they add to the completed model.
    (b) Again, use your discretion when applying these fillet radii to the model.
    (c) These radii are usually R2 or R3, unless a different value is given.

4   As with the P exercises, I have not displayed the assembled models in this chapter, but they are displayed (with the created drawing layouts) on the companion website (www.routledge.com/cw/mcfarlane), in the folder of sample files for Chapter 19.

5   Any animations that I have created can be accessed on the companion website (www.routledge.com/cw/mcfarlane), in the folder of sample files for animations.

Exercise S1: bracket and casting

1   Drawing detail in Third Angle projection of:
   (a)   Part 1: Bracket.
   (b)   Part 2: Casting base – size 100×100 (recommended) $\times$ 24.
   (c)   Others:
   •   M12 Hex Bolt $\times$ 60 (1 of).
   •   M12 Hexagon Nut (1 of).
   •   M12 Washer (2 of).

2   The only stipulation is that, when assembled, centre lines CC and AA have to coincide.

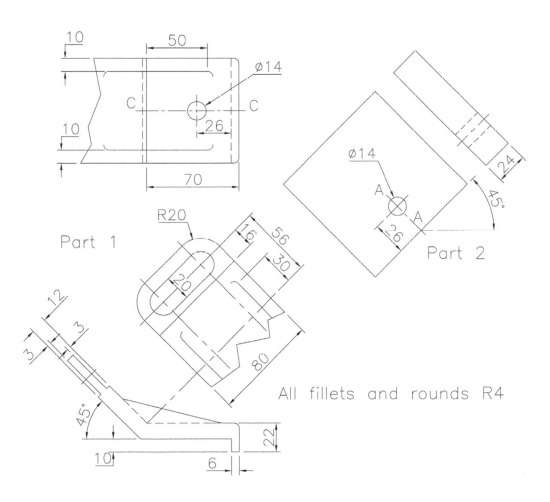

Part 1

Part 2

All fillets and rounds R4

Exercise S2: bracket and base

Third Angle projection drawing detail of the following parts:

1   Part 1: Base.

2   Part 2: Bracket.

3   Others:
   (a) M12-6g Hex Head Bolt, length 2, thread length 24 – Library.
   (b) M12 washer – Library.

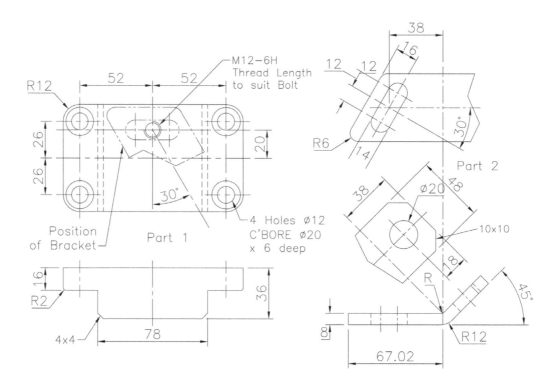

Exercise S3: cam

1   There are six parts to be created for this assembly, displayed in Third Angle projection:
   (a)   Part 1: Cam Base.
   (b)   Part 2: End Cover.
   (c)   Part 3: Case Hardened Mild Steel Camshaft.
   (d)   Part 4: Case Hardened Mild Steel Follower.
   (e)   Part 5: Bronze Follower Bush.
   (f)   Part 6: Bronze Camshaft Bush (2 of).
   (g)   Others: M10-6g Hex Head Screws, 20 lg (3 of).

2   Part 1 notes:
   (a)   Three holes M10-6H equispaced on 84 PCD with 20 min length full thread.
   (b)   Four holes Ø10.
   (c)   Chamfer 2 × 45 degrees at three places.
   (d)   Unspecified radii R3 or to suit.
   (e)   There are several fillet radii to add to this part – your choice.

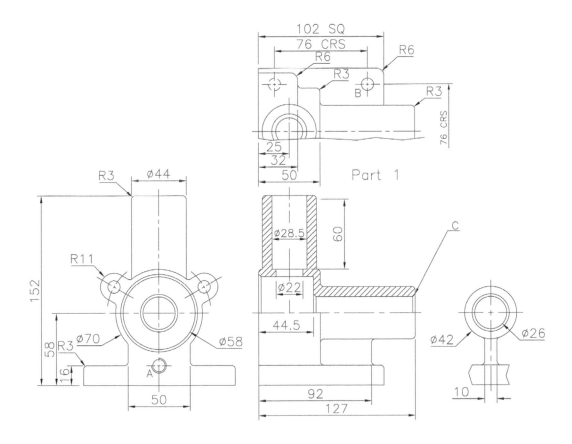

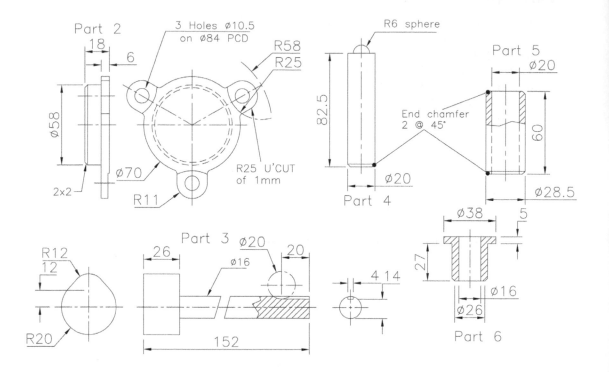

Part 2

18

6

Ø58

2×2

Ø70

Ø11

R11

3 Holes Ø10.5
on Ø84 PCD

R58

R25

R25 U'CUT
of 1mm

R6 sphere

82.5

Ø20

Ø20

Part 4

Part 5

Ø20

End chamfer
2 @ 45°

60

Ø28.5

R12

12

R20

Part 3   Ø20

26

Ø16

20

152

4 14

Ø38

5

27

Ø26

Ø16

Part 6

Exercise S4: belt tension bracket

First Angle projection detail for the following parts of the assembly:

1   Part 1: Bracket.

2   Part 2: Spindle.

3   Part 3: Pulley.

4   Part 4: Collar.

5   Part 5: Handle.

6   Part 6: M12 Washer.

7   Part 7: M12 Hex Nut.

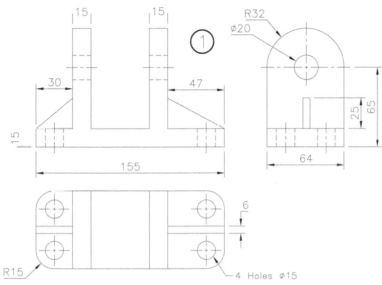

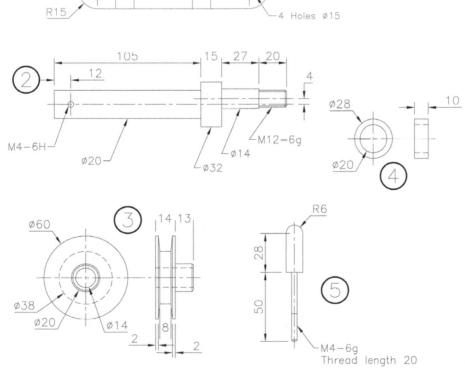

Exercise S5: spindle and holders

A very simple assembly with the two parts detailed below:

1   Part 1: Holder (2 of).

2   Part 2: Spindle.

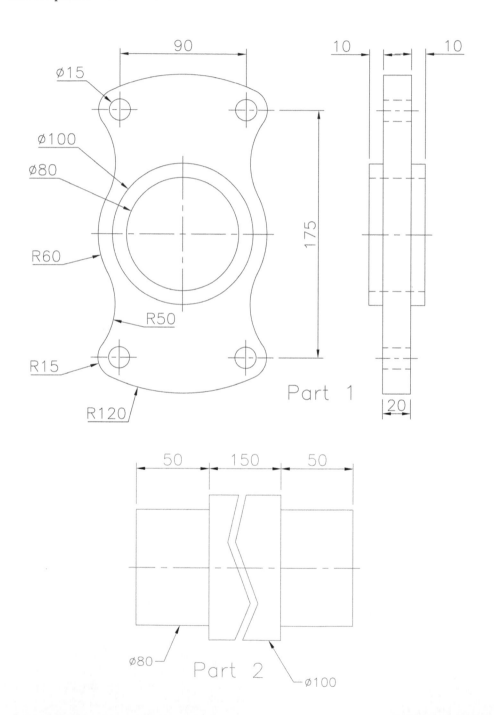

Exercise S6: V–blocks

The parts for the assembly are displayed in First Angle projection:

1   Part 1: Block.

2   Part 2: Spider.

3   Part 3: Screw.

4   Part 4: Tommy Bar.

5   Part 5: Component – user to design to suit the V–Blocks.

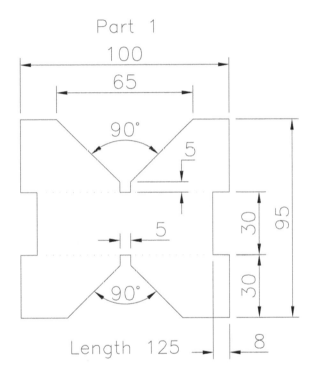

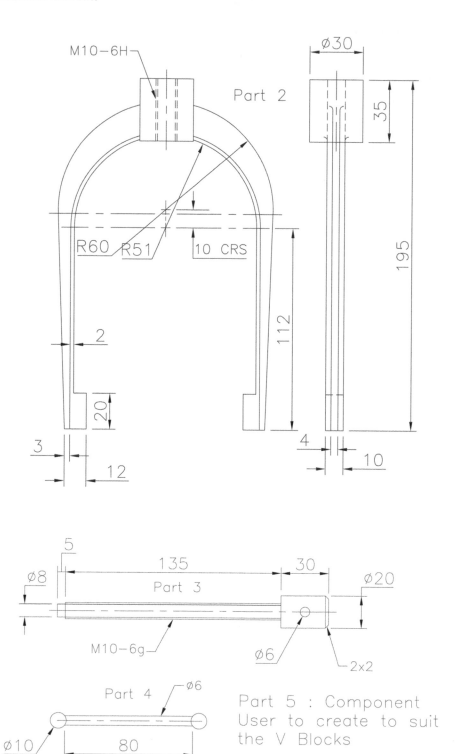

M10−6H

Part 2

⌀30

35

195

R60 R51   10 CRS

112

2

20

3

12

4

10

5

⌀8

135

Part 3

30

⌀20

M10−6g

⌀6

2x2

Part 4   ⌀6

⌀10   80

Part 5 : Component
User to create to suit
the V Blocks

Exercise S7: spinning rig fixture

The views of the parts for this assembly are in Third Angle projection:

1   Part 1: Base.
2   Part 2: Top.
3   Part 3: Lower Long Arms (4 of).
4   Part 4: Upper Short Arms (3 of).
5   Part 5: Lower Short pins – Ø10 with length 42 and 1 × 1 chamfer at ends (4 of).
6   Part 6: Upper Long pins – Ø10 with length 57 and 1 × 1 chamfer at ends (3 of).

Note:

1   The upper arms have to be 30 degrees inclined upwards.
2   The lower arms have to be 45 degrees inclined downwards.

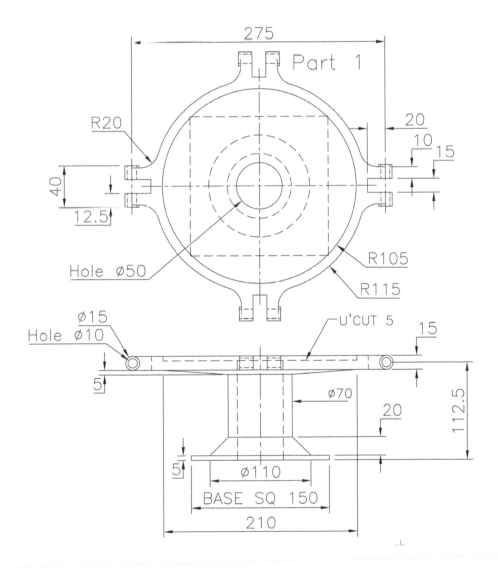

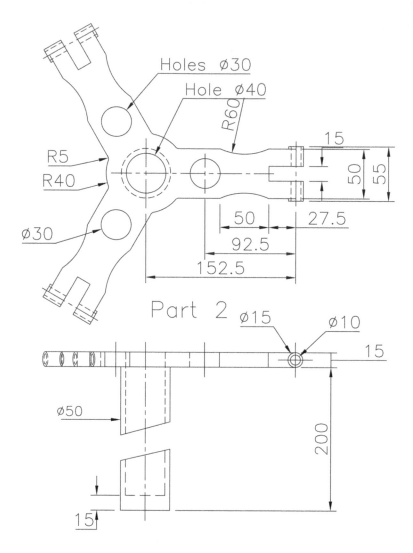

Holes ⌀30
Hole ⌀40
R60
R5
R40
⌀30
15
50
55
50
27.5
92.5
152.5

Part 2   ⌀15   ⌀10
15
⌀50
200
15

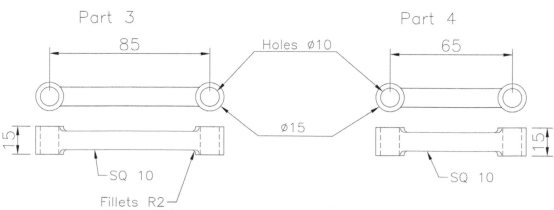

Part 3   Holes ⌀10   Part 4
85   65
15   ⌀15   15
SQ 10
Fillets R2   SQ 10

Exercise S8: lathe steady

Third Angle projection drawings for the following parts:

1   Part 1: Body.
2   Part 2: Body Screw.
3   Part 3: Adjusting Screw (2 of).

4   Part 4: Base Screw (2 of).
5   Part 5: Pin (2 of).
6   Part 6: Holding Screw (2 of).

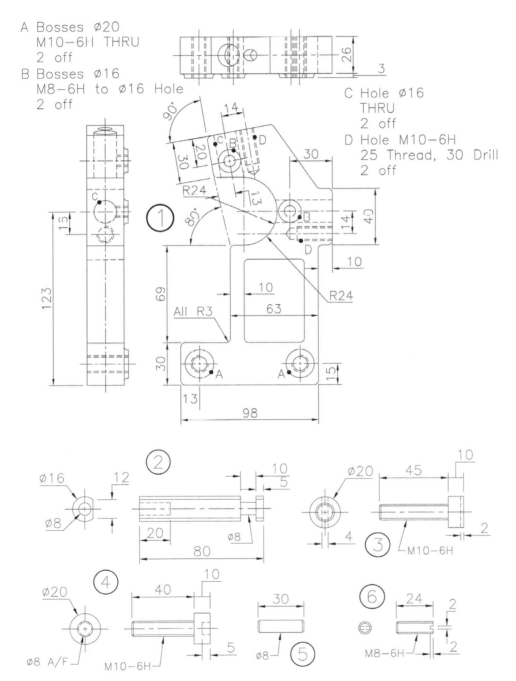

A Bosses ⌀20
   M10−6H THRU
   2 off
B Bosses ⌀16
   M8−6H to ⌀16 Hole
   2 off

C Hole ⌀16
   THRU
   2 off
D Hole M10−6H
   25 Thread, 30 Drill
   2 off

Exercise S9: small lathe tailstock

First Angle projection drawings of the parts:

1   Part 1: Body.

2   Part 2: Hand Wheel.

3   Part 3: Back Plate.

4   Part 4: Sleeve.

5   Part 5: Screw.

6   Part 6: M12-6g Hex Nut (1 of).

7   Part 7: M12 Washer (1 of).

8   Part 8: M6-6g CSK Head screws (4 of).

Note: This was one of the most difficult models in the book for me to create, and I am not convinced I created it correctly.

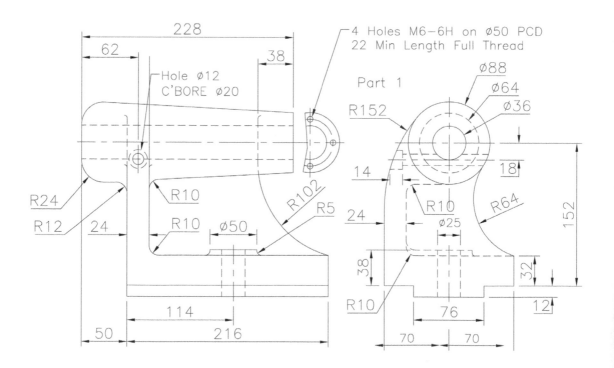

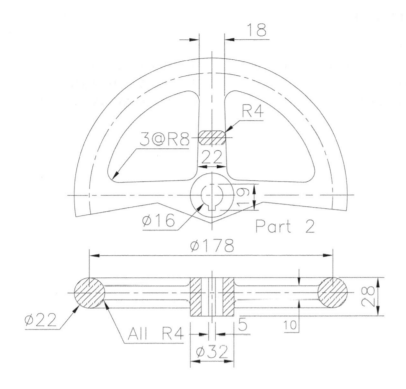

18

R4

3@R8

22

Ø16

19

Part 2

Ø178

Ø22

All R4

5

10

28

Ø32

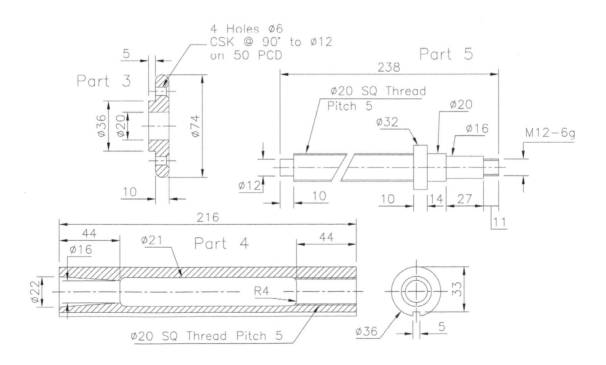

4 Holes Ø6
CSK @ 90° to Ø12
on 50 PCD

Part 3

Part 5

238

5

Ø20 SQ Thread
Pitch 5

Ø20

Ø36

Ø20

Ø74

Ø32

Ø16

M12-6g

10

Ø12

10

10

14

27

11

216

44

44

Ø16

Ø21

Part 4

Ø22

R4

33

Ø20 SQ Thread Pitch 5

Ø36

5

Exercise S10: pulley

Third Angle projection drawings of the following parts:

1   Part 1: Bracket Cast Iron (1 of).
2   Part 2: Pulley Mild Steel (1 of).
3   Part 3: Pin Mild Steel (1 of).

4   Part 4: Bush Bronze (2 of).
5   Part 5: M16 Washer (1 of).
6   Part 6: M16-6H Hex Nut (1 of).

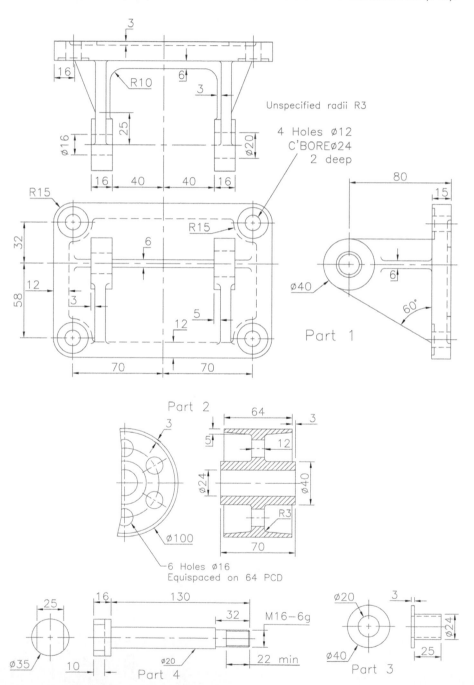

Exercise S11: cylinder relief valve

Third Angle projection drawings of the following parts:

1   Part 1: Body (1 of).
2   Part 2: Valve Seat (1 of).
3   Part 3: Spring Cap (1 of).

4   Part 4: Compression Screw (1 of).
5   Part 5: Valve (1 of).
6   Part 6: Spring (1 of).

Spring details:

1   Inside diameter: 24.
2   Wire diameter: 8.
3   Free length: 165.

4   Working length: to be determined by user.
5   Working coils: 10 Ends ground square to give a 'flat'.

Note: For this assembly, I decided to:

1   display the assembly with part 1 as a 'half-part'; and
2   used the 'complete' part 1 for the drawing layout.

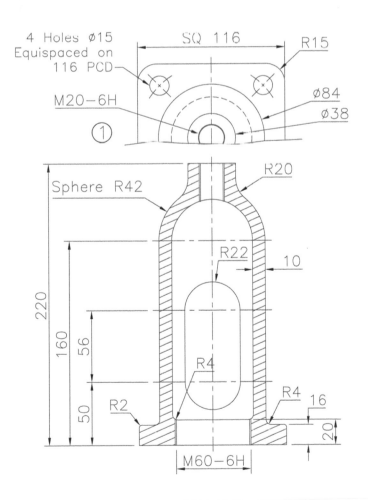

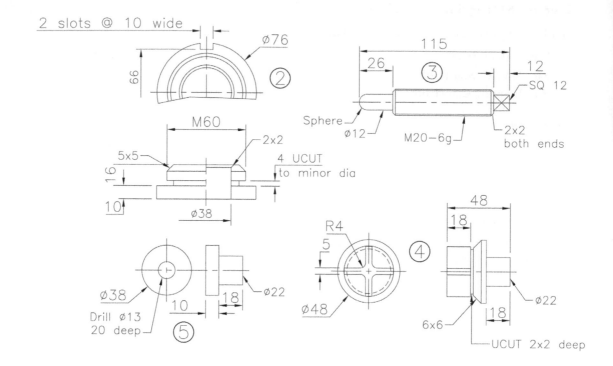

2 slots @ 10 wide

Ø76

②

66

M60

2x2

5x5

4 UCUT
to minor dia

16

10

Ø38

115

26

③

12

SQ 12

Sphere

Ø12

M20—6g

2x2
both ends

R4

5

④

Ø48

48

18

6x6

Ø22

18

UCUT 2x2 deep

Ø38

Drill Ø13
20 deep

⑤

10

18

Ø22

Exercise S12: strut attachment

First Angle projection drawings of the six parts (all mild steel):

1   Part 1: Bracket.

2   Part 2: Fork.

3   Part 3: Bolt.

4   Part 4: Strut End.

5   Part 5: M20-6H Hex Nut.

6   Part 6: M20 Washer.

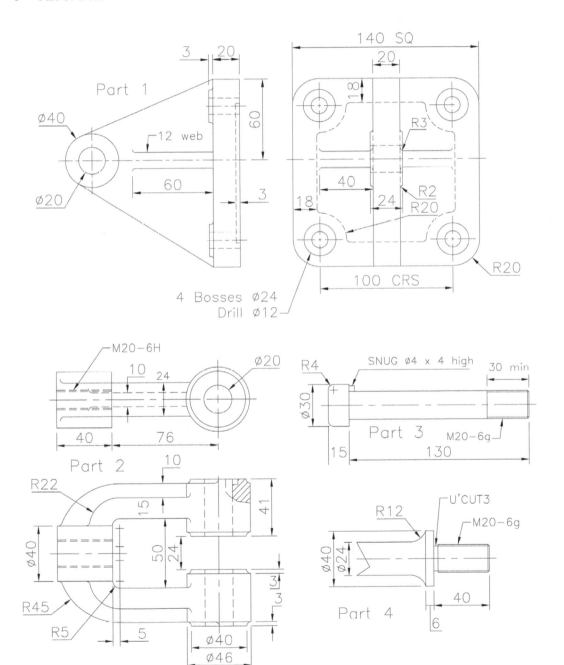

Exercise S13: adjustable spanner

First Angle projection drawings of the six parts:

1   Part 1: Fixed Jaw.
2   Part 2: Sliding Jaw.
3   Part 3: Knurled Nut.
4   Part 4: Sliding Bracket.
5   Part 5: Fixed Bracket.
6   Part 6: Rivets (4 of).

Note: The fixed and sliding jaws have to be assembled 25mm apart.

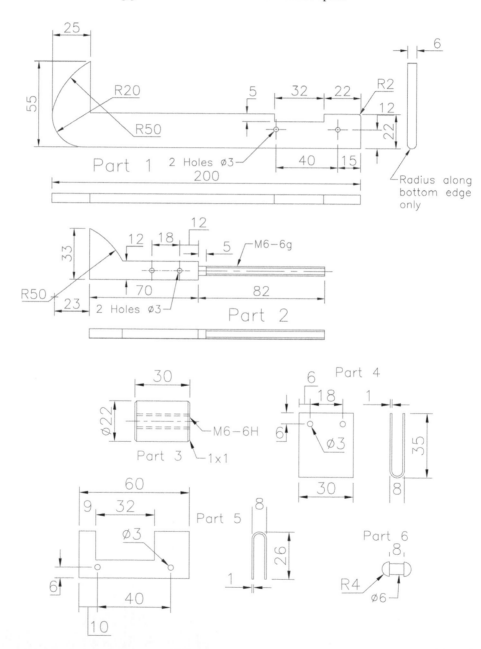

Exercise S14: trolley wheel

First Angle projection drawings of the parts:

1   Part 1: Wheel Bracket.
2   Part 2: Wheel.
3   Part 3: Pivot.
4   Part 4: M10 Washer (3 of).

5   Part 5: M10-6H Hex Nut.
6   Part 6: Spindle.
7   Part 7: Wheel Bush (2 of).
8   Part 8: Pin Ø3 × 14 (2 of).

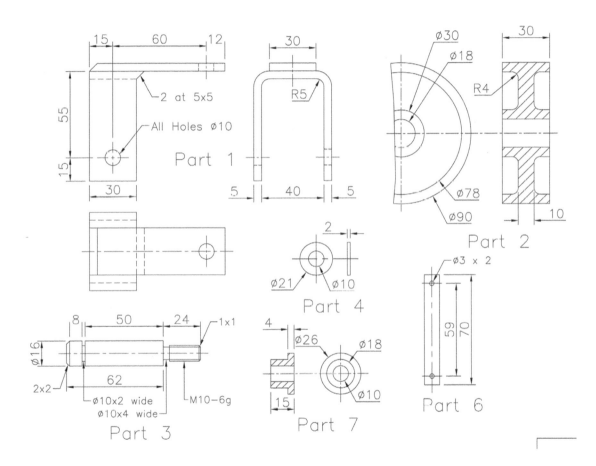

Exercise S15: lever

Three parts in First Angle projection for this assembly:

1   Part 1: Pivot Base.
2   Part 2: Lever.
3   Part 3: Pivot.

Note:

1   Fillet radii at user discretion.
2   The base and pivot have to be inclined at 15 degrees to each other.

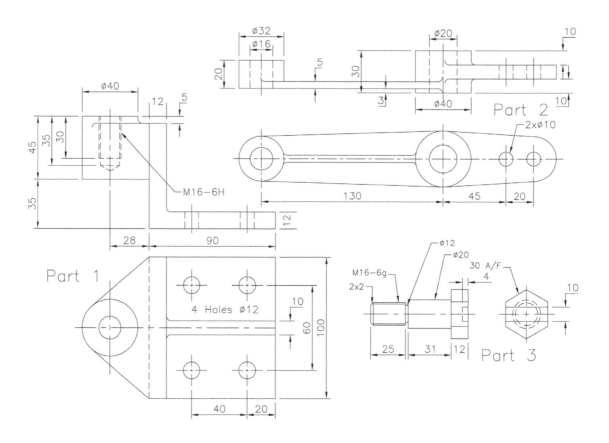

Exercise S16: guide pulley

The parts drawing information for this assembly (in First Angle projection) are:

1 Part 1: Guide Bracket.
2 Part 2: Pulley.
3 Part 3: Clamp Screw.
4 Part 4: Spindle.
5 Part 5: Retaining Bracket.

6 Part 6: M12-6g Hex Nut.
7 Part 7: M12 Washer.
8 Part 8: M6-6g Slotted Screw.
9 Part 9: Bush.

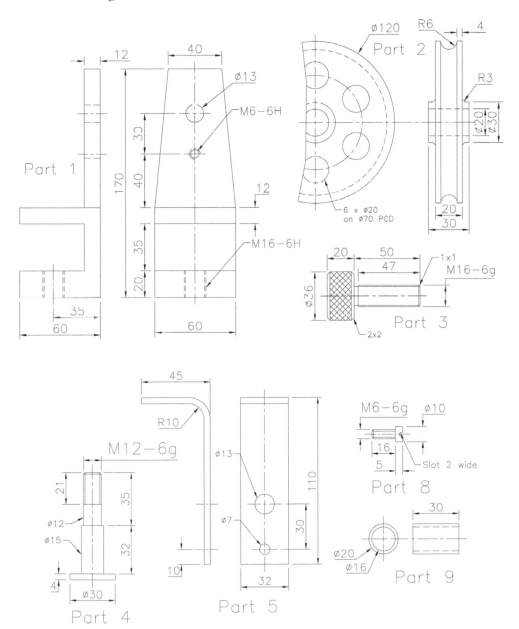

Exercise S17: gear change

The parts for this assembly (in First Angle projection) are:

1   Part 1: Elbow Casting.
2   Part 2: Control Knob.
3   Part 3: Control Rod.

4   Part 4: Pivot.
5   Part 5: M16-6H Locknut.
6   Part 6: M16 Washer.

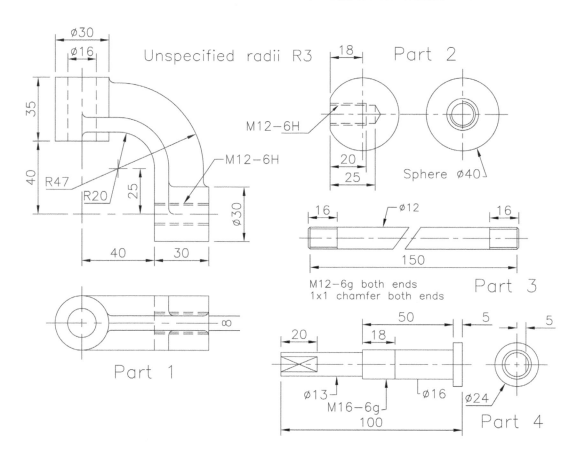

Exercise S18: clamp

Five parts (displayed in First Angle projection) for this assembly:

1   Part 1: Clamp Base.
2   Part 2: Sliding Jaw.
3   Part 3: Clamp Screw.

4   Part 4: M10-6H Hex Nut.
5   Part 5: M10 Washer.

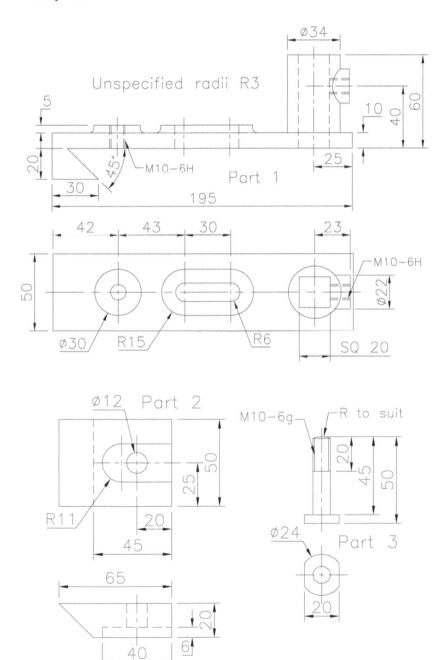

Exercise S19: hook and pulley

The parts for this First Angle projection assembly are:

1 Part 1: Hook.
2 Part 2: Pulley.
3 Part 3: Side Link (2 of).
4 Part 4: Support Block.
5 Part 5: Spindle.

6 Part 6: M16-6H Hex Nut.
7 Part 7: M16 (or appropriate) Locknut.
8 Part 8: M14 Washer (4 of).
9 Part 9: M16 Washer (1 of).
10 Part 10: Pin (4 of, Ø3 × 20).

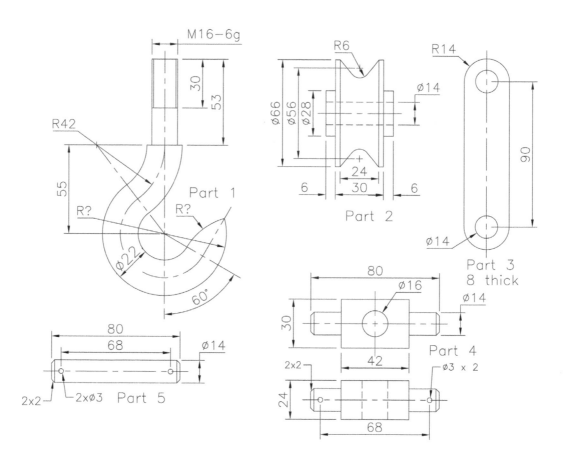

Exercise S20: roller support

The parts for this First Angle projection assembly are:

1 Part 1: Roller Base.

2 Part 2: Nylon Roller.

3 Part 3: Roller Spindle.

4 Part 4: Washer.

5 Part 5: Pin − Ø3 × 28.

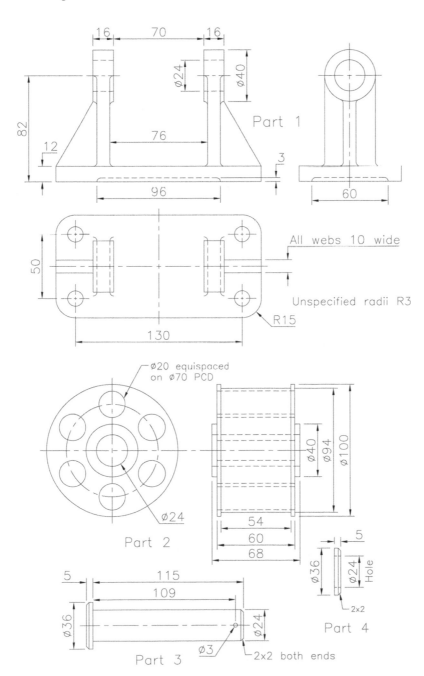

Exercise S21: cylindrical component clamp

The four parts in this assembly are:

1   Part 1: Clamp.
2   Part 2: V Block.
3   Part 3: Clamp Screw.
4   Part 4: Component – own design.

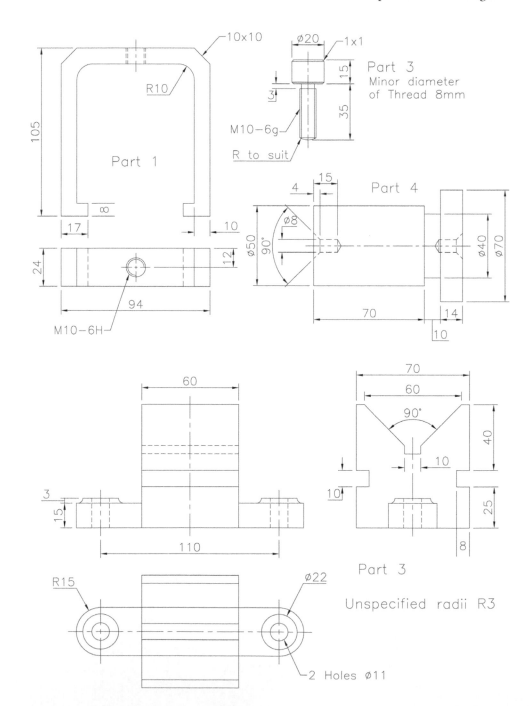

Exercise S22: non-return valve

The parts for this assembly are:

1 Part 1: Valve Body.

2 Part 2: Cover.

3 Part 3: Valve.

4 Part 4: Valve Seat.

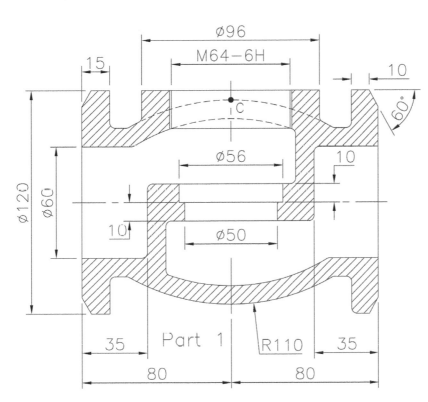

Notes: 1 R110 centre point C is 5mm from top horiz line
2 Wall thickness is uniformly 10mm
3 All unspecified radii R3

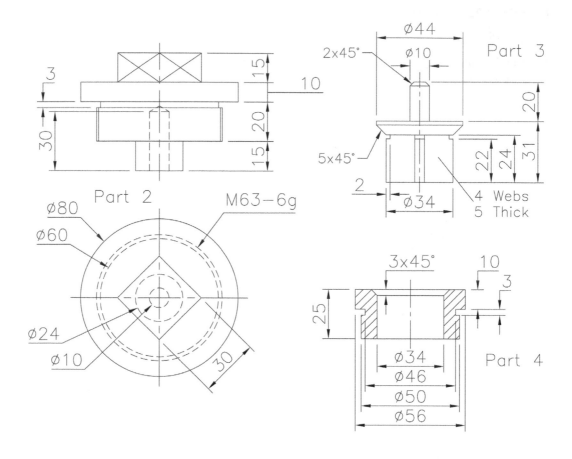

Part 2

Ø80
Ø60
Ø24
Ø10
M63-6g
30
3
30
15
20
15
10

Part 3

Ø44
Ø10
2x45°
5x45°
2
Ø34
20
22
24
31
4 Webs
5 Thick

Part 4

3x45°
10
3
25
Ø34
Ø46
Ø50
Ø56

Exercise S23: machine vice

Several parts for this assembly:

1 Part 1: Base.

2 Part 2: Fixed Block.

3 Part 3: Sliding Jaw.

4 Part 4: Fixed Jaw.

5 Part 5: Spindle.

6 Part 6: Side Bracket (2 of).

7 Part 7: M8-6H Screw (4 of).

8 Part 8: M5-6H Screw (4 of).

9 Part 9: M3-6H Screw (2 of).

10 Part 10: Component – $\varnothing 20 \times 120$ length.

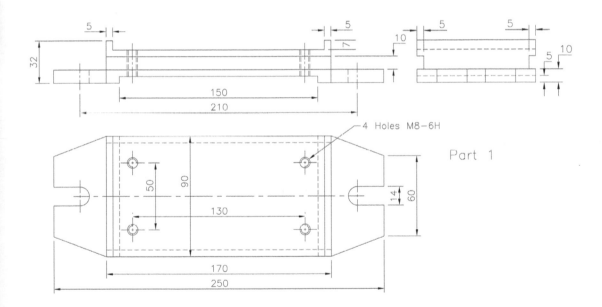

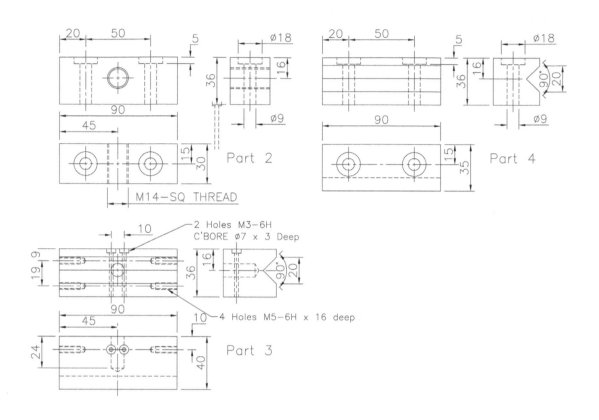

Part 2

M14—SQ THREAD

Part 4

2 Holes M3—6H
C'BORE ⌀7 x 3 Deep

4 Holes M5—6H x 16 deep

Part 3

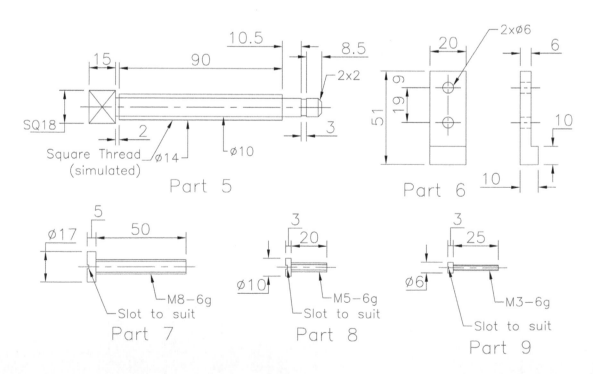

Square Thread
(simulated) ⌀14

Part 5

SQ18

2x2

⌀10

2x⌀6

Part 6

Part 7

⌀17

Slot to suit

M8—6g

Part 8

⌀10

Slot to suit

M5—6g

Part 9

⌀6

Slot to suit

M3—6g

Exercise S24: safety valve

In Third Angle projection, the parts for this assembly are:

1   Part 1: Valve Base.

2   Part 2: Valve.

3   Part 3: Valve Spring.

4   Part 4: Valve Body.

5   Part 5: Valve Cover.

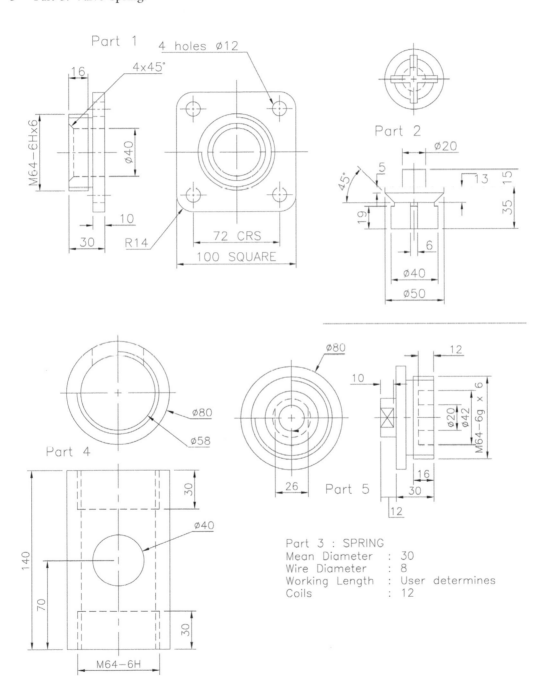

Exercise S25: work fixture

Several parts for this assembly:

1   Part 1: Base.
2   Part 2: Vee Block.
3   Part 3: Clamp Screw.
4   Part 4: Pad.
5   Part 5: Screw Support.
6   Part 6: Knob.

7   Part 7: M6-6H Screw (2 of).
8   Part 8: M10-6H Screw (2 of).
9   Part 9: Stop.
10 Part 10: M12 Locknut.
11 Part 11: Work Feature ⌀50 × 50 length.

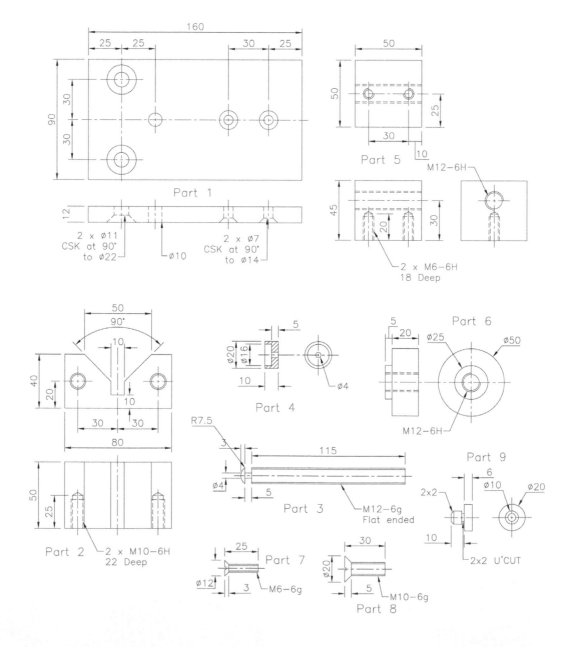

T Projects

1   This chapter has a series of exercises/tasks that I have called projects, and:
    (a) they are all assemblies; and
    (b) they are not any easier or harder than some of the exercises in the chapters that you have completed, but they may (perhaps) take longer to complete.

2   I have tried to add variety to these projects, so I hope you enjoy completing them.

3   They have all been completed using Inventor® and the 2D drawings use AutoCAD.

4   The 10 projects are:
    (a) Push and Go Trike.
    (b) Wall Clock.
    (c) Chess Set.
    (d) Hand Winch.
    (e) Newton's Cradle.
    (f) Turbine Disc.
    (g) Compressor Disc.
    (h) Oscillating Steam Engine.
    (j) Plastic Cube Puzzle.
    (j) Möbius Strip.

5   With each project, you have to:
    (a) read the write-up and review the drawing information given;
    (b) create models of each part required for the assembly;
    (c) create an assembly using the part models; and
    (d) complete the requirements for the project, which may be a drawing layout, an exploded view, a presentation file and/or an animation.

6   The project models and layouts are all displayed on the companion website (www.routledge.com/cw/mcfarlane), in the folder of sample files for Chapter 20.

7   A completed (monochrome) model will be displayed, which may assist you to complete the assembly.

Project 1: push and go trike

1   A young engineering designer working for a large toy manufacturer has been tasked to design a push and go trike for children.

2   The only limitations placed upon the designer are that the completed trike is to be colourful, sturdy and steerable.

3   Using the basic drawing information from the designer:
    (a)   create each part, adding colour as appropriate;
    (b)   create an assembly of the trike; and
    (c)   produce a drawing layout to your own specification.

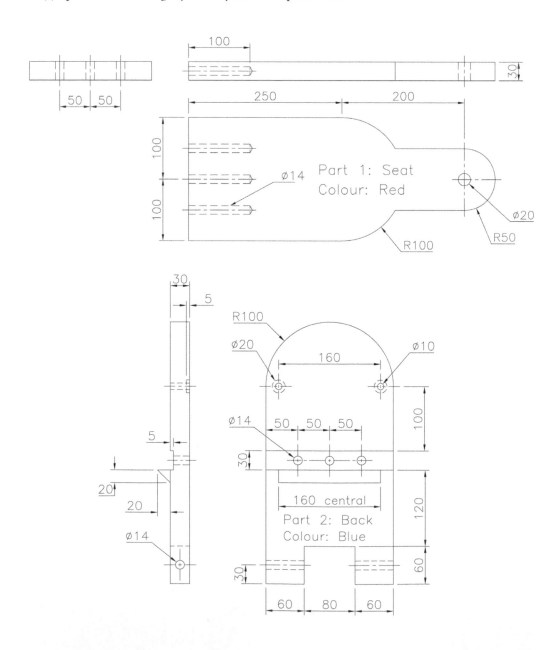

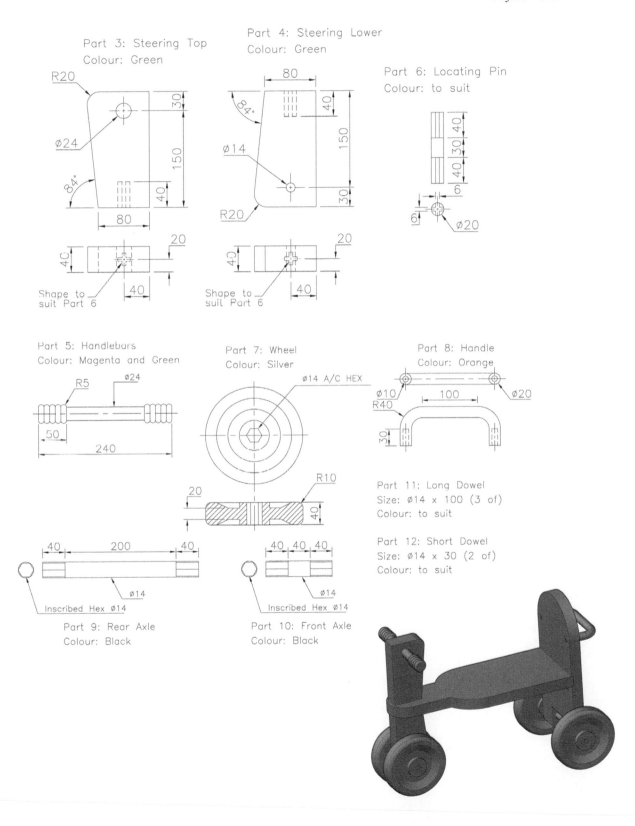

Part 3: Steering Top
Colour: Green

R20

Ø24

84°

30

150

40

80

20

40

4C

Shape to
suit Part 6

Part 4: Steering Lower
Colour: Green

80

84°

40

150

30

Ø14

R20

20

40

4C

Shape to
suit Part 6

Part 6: Locating Pin
Colour: to suit

40
30
40

6

6

Ø20

Part 5: Handlebars
Colour: Magenta and Green

R5   Ø24

50

240

Part 7: Wheel
Colour: Silver

Ø14 A/C HEX

20

R10

40

Part 8: Handle
Colour: Orange

Ø10
R40

100

Ø20

30

Part 11: Long Dowel
Size: Ø14 x 100 (3 of)
Colour: to suit

Part 12: Short Dowel
Size: Ø14 x 30 (2 of)
Colour: to suit

40   200   40

Ø14

Inscribed Hex Ø14

Part 9: Rear Axle
Colour: Black

40   40   40

Ø14

Inscribed Hex Ø14

Part 10: Front Axle
Colour: Black

Project 2: wall clock

1   I used this 'exercise' in my AutoCAD modelling books when rendering was first introduced by Autodesk®. It was an interesting model then, as it incorporated materials, lighting and shadow effects.

2   As a feature-based model, it is still interesting and allows the user various design options with the clock face numerals, the hands and the colours/materials to use for the parts.

3   For this project, you have to create:
(a) the various named parts from the drawing details displayed, attaching materials of your choice;
(b) an assembly of the model;
(c) an exploded view of the assembly;
(d) a drawing layout to your own specification;
(e) a presentation file to your own specification; and
(f) an animation of the pendulum and hands to your specification.

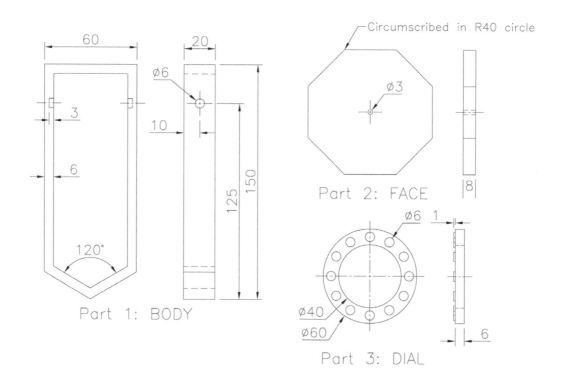

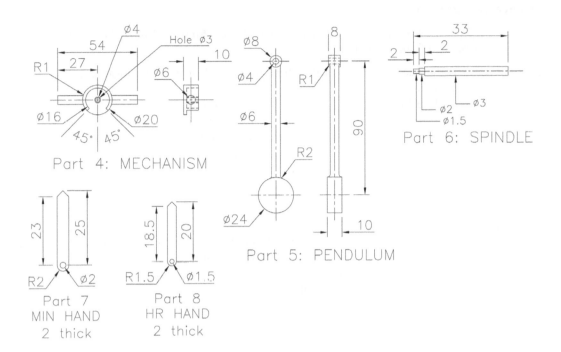

Part 4: MECHANISM

Part 5: PENDULUM

Part 6: SPINDLE

Part 7
MIN HAND
2 thick

Part 8
HR HAND
2 thick

Project 3: chess set

1   This project is another exercise that was included in my AutoCAD modelling books, and is very suitable as a feature-based project.

2   The user has the freedom to design the chess pieces to their own specifications, within the following limitations:
   (a)   The chess board has to be created from 80 × 80 × 10 cuboids.
   (b)   The chess pieces should have a suitable base radius but no less than 50mm diameter.
   (c)   As a suggestion, I will display my basic drawings for five pieces with the model I created, but, as stated, these pieces should be your own design.
   (d)   The dimensions displayed are only given as a guide.
   (e)   The knight has been left for you to design completely, but I have displayed the model I created.

3   You have to create:
   (a)   the various chess pieces from the drawing details displayed (or your own design), attaching two materials/colours to each piece and saving with appropriate names;
   (b)   an assembly of the chess board with all pieces correctly positioned;
   (c)   a drawing layout to your own specification;
   (d)   a presentation file to your own specification; and
   (e)   an animation of a 'partial' game (within limitations, of course).

PAWN

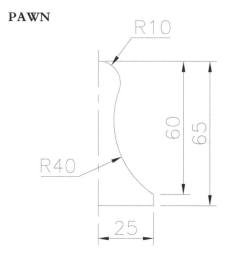

ROOK

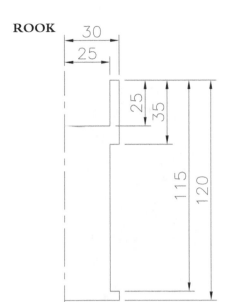

BISHOP

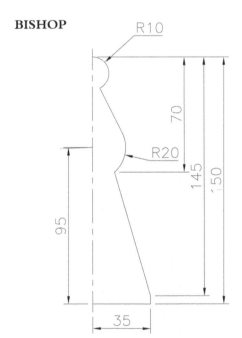

QUEEN

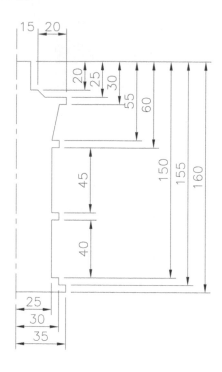

KING

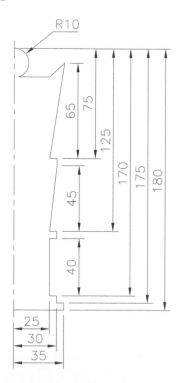

KNIGHT

COMPLETE CHESS SET

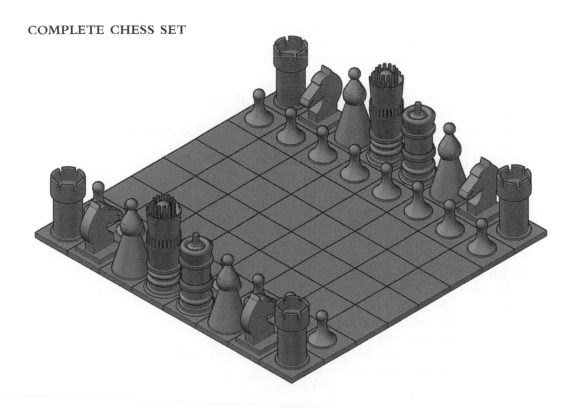

Project 4: hand winch

1   This is a project that was given to former HNC CADD students when I worked in further education.

2   It was considered a very good project as it tested the students' ability with the basic concepts of feature-based modelling.

3   I have left the write-up and requirements as presented to the students.

4   *Remit*: A hand wrench is to be created as an assembly from the following parts:

Item	Quantity	Part name	Material
1	1	BASE_PLATE-MOD1	Metal-Steel (Polished)
2	2	SIDE_PLATE-MOD1	Metal-AL-6061 (Flat)
3	1	COG_WHEEL-MK3	Metal-Titanium (Polished)
4	1	SOCKET_MOD3	Metal-AL-6061 (Polished)
5	1	HANDLE-MK2	Rubber Blue
6	2	BUSH-MOD1	Metal-Brass
7	2	PIN	Nickel (Bright)
8	4	M4 slotted Head Bolt, length 20	Mild Steel
9	4	M4 Hex Nut	Mild Steel
10	4	M4 Washer	Mild Steel

5   (a)   Each of the seven main parts in the assembly will be displayed as Third Angle projection drawings with the dimensions required to create the model.
      (b)   As usual, use your discretion if you are not too sure about a given size.

6   *Requirements*
      (a)   Create a part model for items 1–7 necessary for the assembly, saving as usual.
      (b)   Create a drawing layout for each part to display:
      •   the three basic orthographic views in *either First or Third Angle* projection;
      •   a 3D view;
      •   at least one section and one detail view;
      •   suitable dimensions;
      •   the drawing layout sheet properly completed; and
      •   no drawing layout required for part 7.
      (c)   Create and display an assembly of the hand winch with all items included.
      (d)   Create and display an exploded assembly as a drawing layout with all items identified and a parts list included.
      (e)   For the assembly, produce the following drawing layouts:
      •   the basic three orthogonal views (without hidden detail) and a 3D view;
      •   the orthographic views with hidden detail, with dimensions;
      •   a display with some section views; and
      •   a display with some detail views.
      (f)   a presentation/animation file for the complete assembly; and
      (g)   an Animation file for the rotating parts only.

Part 1: **BASE_PLATE–MOD1** drawing information with the part model display.

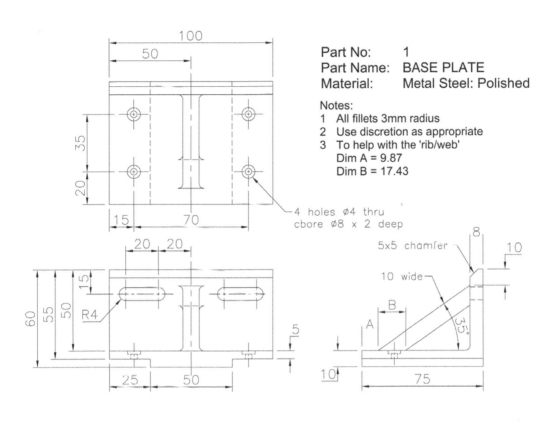

Part No: 1
Part Name: BASE PLATE
Material: Metal Steel: Polished

Notes:
1 All fillets 3mm radius
2 Use discretion as appropriate
3 To help with the 'rib/web'
   Dim A = 9.87
   Dim B = 17.43

4 holes ⌀4 thru
cbore ⌀8 x 2 deep

5x5 chamfer

10 wide

Part 2: **SIDE_PLATE-MOD1** drawing information with the part model display.

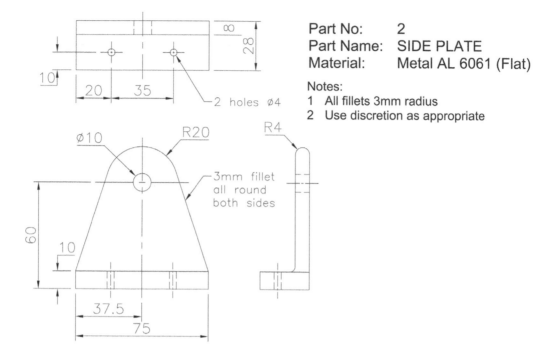

Part No: 2
Part Name: SIDE PLATE
Material: Metal AL 6061 (Flat)

Notes:
1   All fillets 3mm radius
2   Use discretion as appropriate

Part 3: **COG_WHEEL-MK3** drawing information with the part model display.

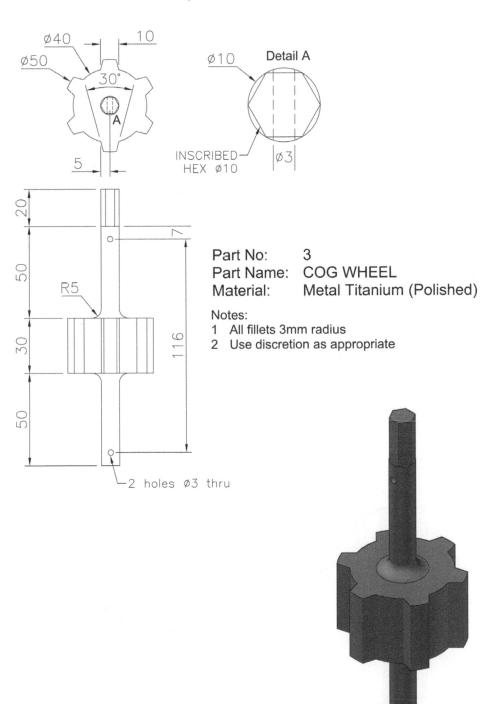

Ø40

10

Ø50

30°

A

5

Detail A

Ø10

INSCRIBED
HEX Ø10

Ø3

20

7

50

R5

30

116

50

2 holes Ø3 thru

Part No: 3
Part Name: COG WHEEL
Material: Metal Titanium (Polished)

Notes:
1   All fillets 3mm radius
2   Use discretion as appropriate

Part 4: **SOCKET–MOD3** drawing information with the part model display.

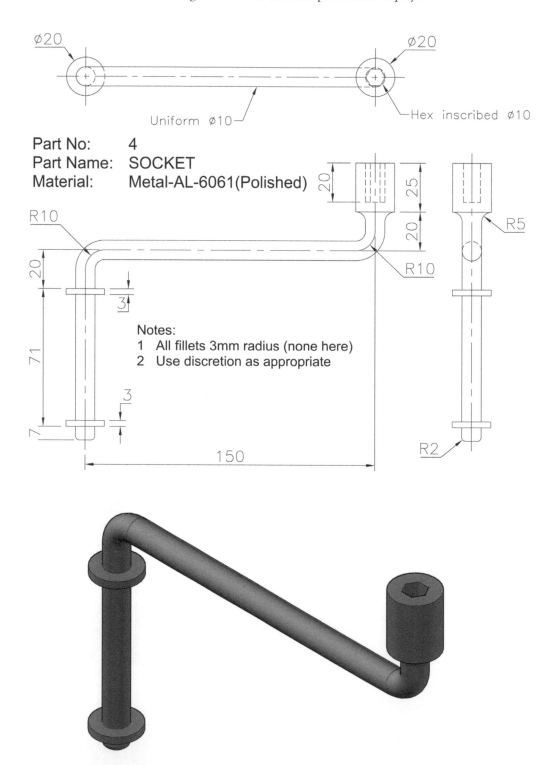

∅20   ∅20

Uniform ∅10   Hex inscribed ∅10

Part No: 4
Part Name: SOCKET
Material: Metal-AL-6061(Polished)

R10

20

71

7

R10

R5

R2

Notes:
1   All fillets 3mm radius (none here)
2   Use discretion as appropriate

20

25

20

3

3

150

Part 5: **HANDLE-MK2** drawing information with the part model display.

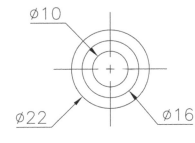

Ø10

Ø22

Ø16

Part No: 5
Part Name: HANDLE
Material: Rubber

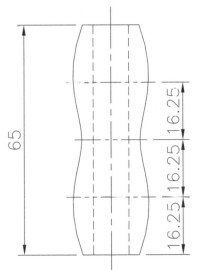

65

16.25 16.25 16.25 16.25

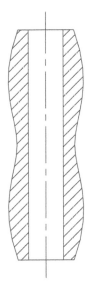

Notes:
1   All fillets 3mm radius (none here)
2   Use discretion as appropriate

Part 6: **BUSH–MOD1** drawing information with the part model display.

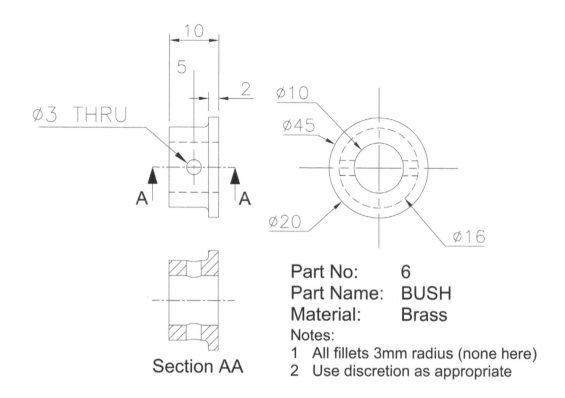

Part No: 6
Part Name: BUSH
Material: Brass
Notes:
1 All fillets 3mm radius (none here)
2 Use discretion as appropriate

Section AA

Part 7: **PIN** drawing information with the part model display.

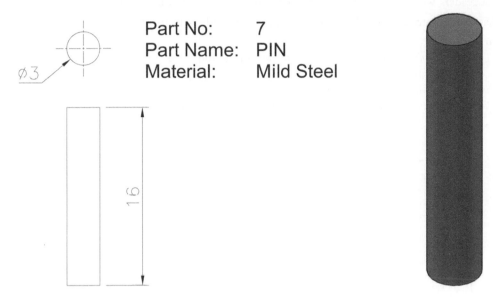

Part No: 7
Part Name: PIN
Material: Mild Steel

Assembly

Project 5: Newton's cradle

1 Most readers will have heard about and probably used a Newton's cradle.

2 It has always intrigued me, hence my decision to include it in this book.

3 You have to:
   (a) use the drawing details below to create each of the four parts (numbered) – the dimensions displayed can be modified by the user as required;
   (b) create an assembly of the cradle; and
   (c) create a presentation/animation for motion of the end spheres.

4 So really, quite a simple project to complete.

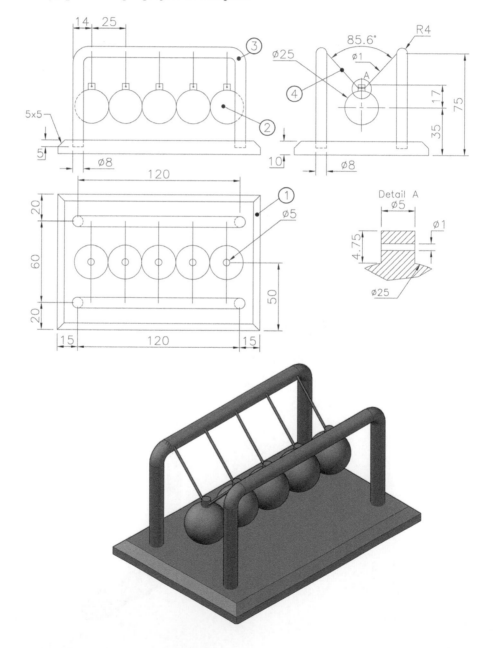

Project 6: turbine disc

1   You are now a design engineer with a well-known international company who manufacture aero engines.

2   (a) You have been tasked by your section leader to create a turbine disc model, but there is a slight problem.
    (b) The only information you have is an old drawing of a turbine blade, which is displayed below.

3   Using your design ability, create:
    (a) a model of the turbine disc using the blade details given – the disc size and modification of the blade details (as required) have been left to your discretion;
    (b) a drawing layout to the section leader specifications; and
    (c) a presentation and animation file of the turbine disc rotation.

4   *Note*:
    (a) You may need to calculate some additional dimensions to complete this model, especially angles.
    (b) The 'turbine blade' has already been created as Excercise O38 during Chapter 15.

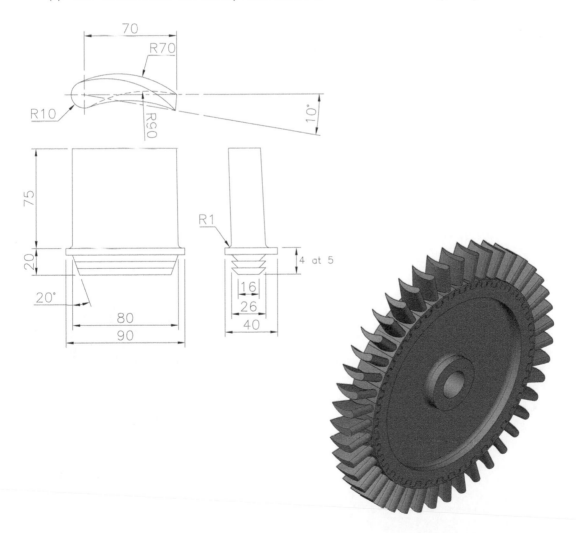

Project 7: compressor disc

1   With your turbine disc model success, the compressor section has asked for your services to complete a similar exercise.

2   They also have an old drawing of a compressor blade (below), and would like you to create:
    (a)   a model of a compressor disc using the blade details given;
    (b)   as with the turbine disc, modification of the blade details (as required) has been left to your discretion, although some calculations may be necessary;
    (c)   a drawing layout to the compressor section leader specifications; and
    (d)   a presentation and animation file of the compressor disc rotation.

3   *Note*: The compressor disc has three parts:
    (a)   the blade – already created as Excercise O39 during Chapter 15;
    (b)   the disc – which you have to design; and
    (c)   a pin – $\varnothing16 \times 140$, which 'holds' the blade to the disc.

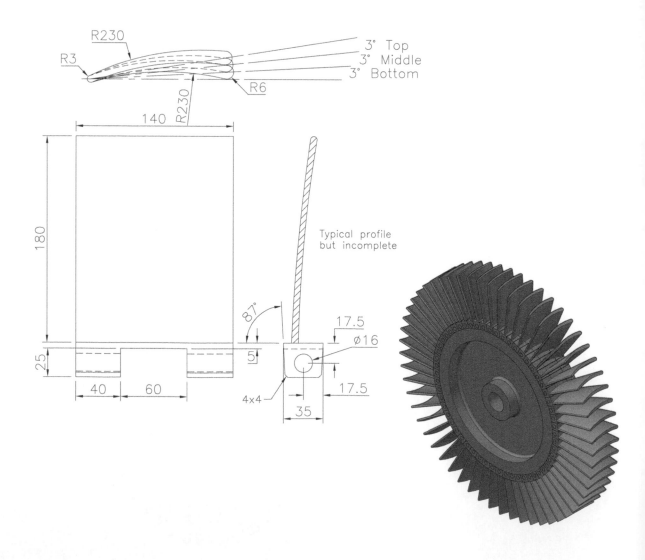

Project 8: oscillating steam engine

1 While thinking up ideas for projects to include in the book, I was 'trawling' through some old CAD drawing material and came across 10 A4 stapled drawings of a 'simple oscillating steam engine', as well as my AutoCAD Solid model of the parts.

2 (a) I decided that this would be a very good project for feature-based modelling.
   (b) This is not an easy project, and requires some thought and perseverance.

3 The drawings (and dimensions) will be displayed as I have them (i.e. without part names, only boxed part numbers), and there are 17 parts required for an assembly.

4 The drawings are basically 'instructions' for the construction of the steam engine.

5 *Project requirements*
   (a) Create models of the 17 parts and save as appropriate. I created the parts with varying materials and colour effect only to assist with the assembly. Note that you may require your discretion with sizes.
   (b) Create an assembly of the 17 parts, which also requires the following Library parts:
   - M6 slot head screw × 20 (or suitable) length (2 of);
   - M3 hex nut; and
   - ∅3 grub screw × 10 (or suitable) length.
   (c) Create the following drawing layouts:
   - First Angle projection basic layout without hidden detail and with dimensions added;
   - Third Angle projection layout with hidden detail included;
   - section views with parts identified and a parts list; and
   - some detail views.
   (d) Create an exploded view of the assembly.
   (e) Create an animated movie file of your choice.
   (f) Obtain the assembly properties.

6 *The assembly*
   (a) I have included a monochrome model of the assembly at the start of the project.
   (b) This is to assist you when creating the various individual parts, as well as the assembly.

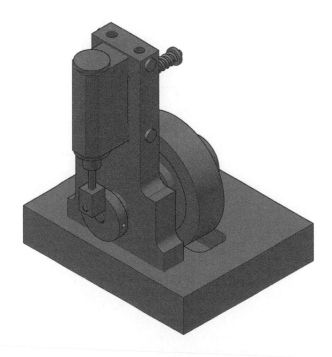

The drawing information for creating the model parts

Part 1

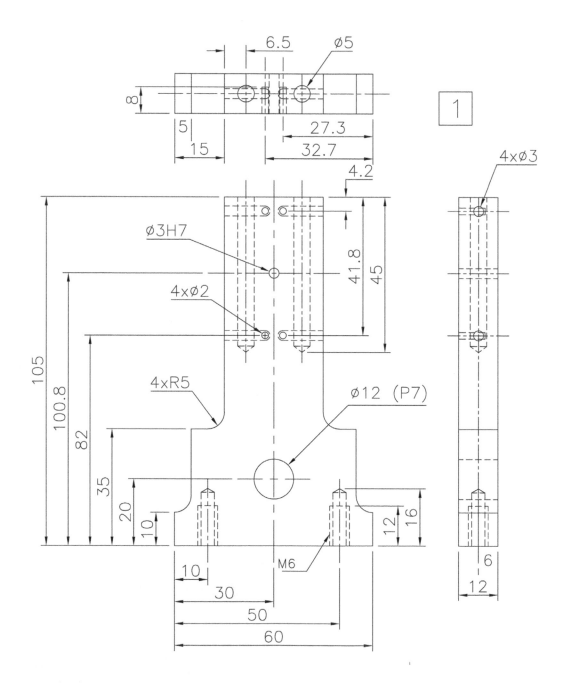

Parts 2, 3 and 6

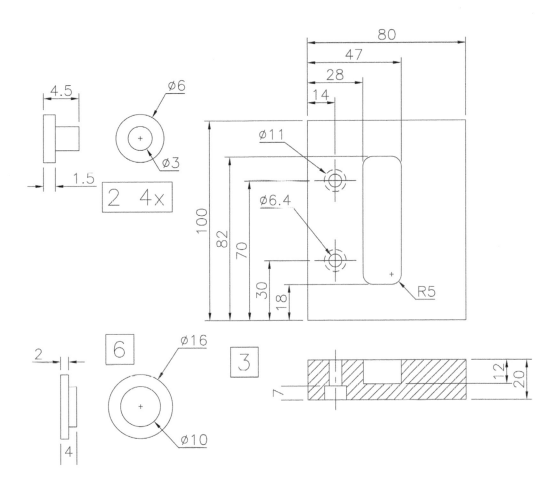

Part 4

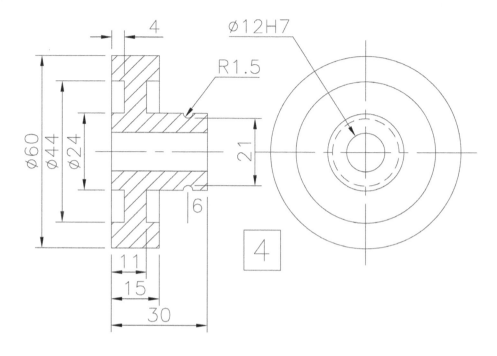

Part 5

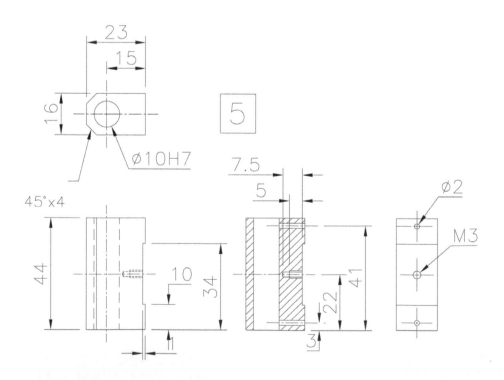

Parts 7, 8, 9 and 10

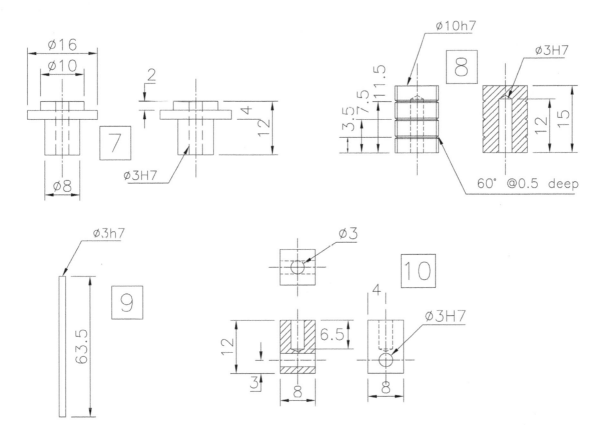

Parts 7, 8, 9 and 10 ASSEMBLY

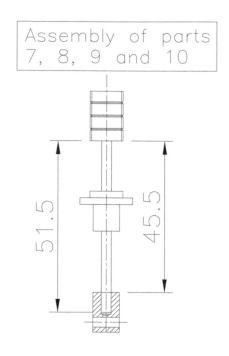

Parts 11, 12, 13 and 14

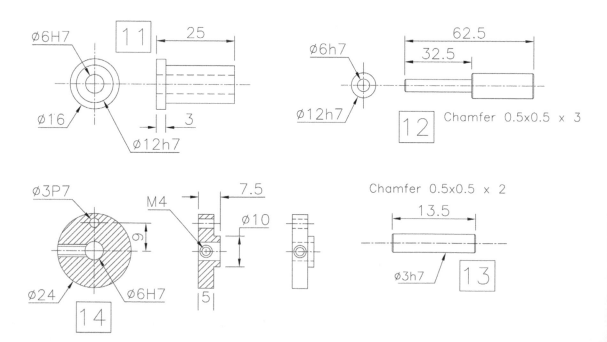

Parts 15, 16 and 17

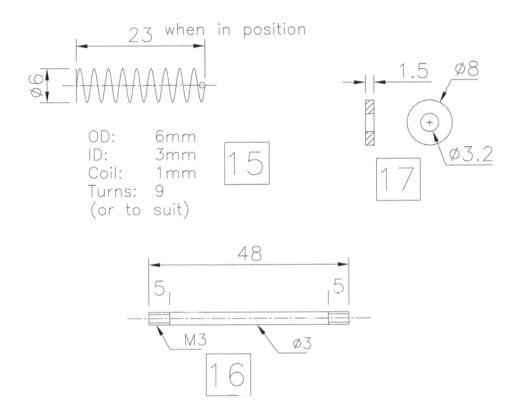

Author's comment on the assembly

1   Some of the assembly constraints are 'tricky', especially part 8, which is 'hidden' inside part 5.

2   I 'assigned' glass material to part 5, to enable me to observe the motion of the piston arrangement inside the cylinder.

3   When assembled, the piston arrangement should move up and down inside the part 5 cylinder, as part 14 is drag rotated and the cylinder itself should oscillate about the Ø3 hole in the back surface – hence the name.

Project 9: plastic cube puzzle

1   A plastic cube (overall size 24 × 24 × 24) puzzle consists of six coloured parts.

2   An isometric view of the cube showing three of the faces is displayed below.

3   You have to create:
   (a) the six parts using the drawing information given;
   (b) an assembly of the cube;
   (c) a drawing layout to display the six faces of the cube and any other details of your choice;
   (d) an exploded view;
   (e) check for interference; and
   (f) an animation of the six parts being assembled/'disassembled'.

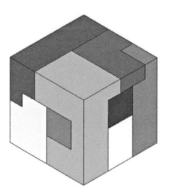

Author note for assembly

1   Select the green part 6 as the ground/anchor piece.

2   The other five pieces all 'slide' into place about the green part and each other.

3   Assemble the parts in 'reverse order' (i.e. part 5, then part 4, etc.).

Part 1: **WHITE**

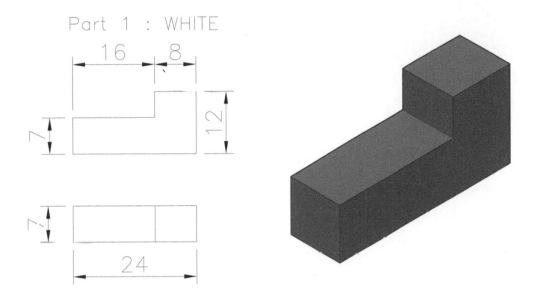

Part 2: **YELLOW**

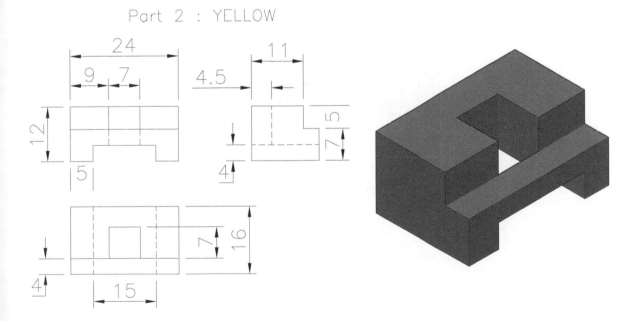

Part 3: BLUE

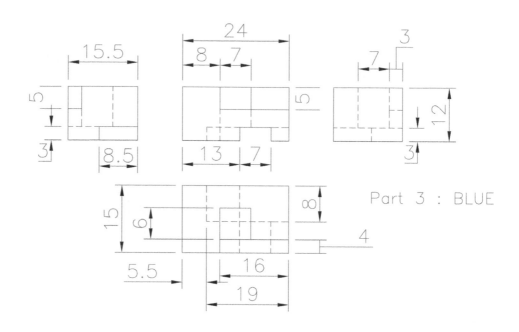

Part 3 : BLUE

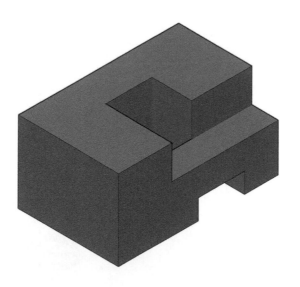

Part 4: ORANGE

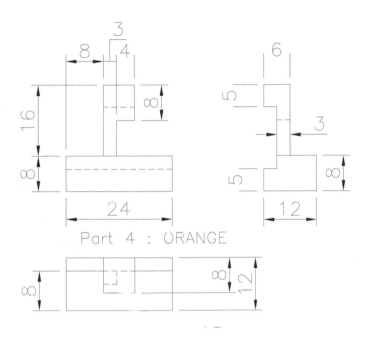

Part 4 : ORANGE

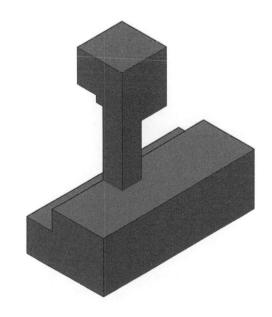

Part 5: **PINK**

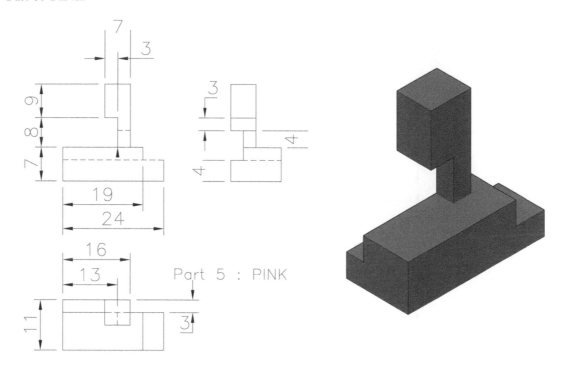

Part 5 : PINK

Part 6: **GREEN**

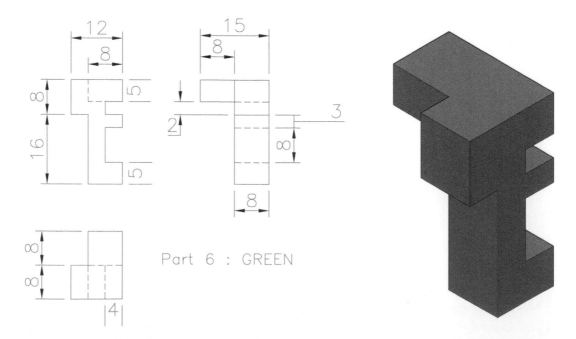

Part 6 : GREEN

Interference

1 On checking for interference, I found that there were three 'clashes', as displayed in the wire frame display of the assembly below, and the table indicated these as:

(a) blue with white;

(b) orange with pink; and

(c) orange with green.

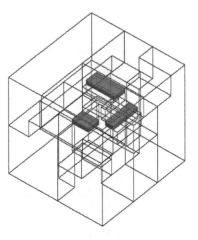

2 As this indicated that I had created the named parts wrongly, I then edited these parts until there was no interference.

3 I realise that I should have modified the drawing information so that there were no clashes in the assembly, but decided to leave the drawing information as given, as this would allow the user to detect interference and modify the various parts, should they want to.

4 The modifications are actually very simple and should present no problem to the reader.

5 The result of editing the parts:

Interference information fron Inventor 7

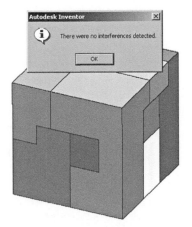

Project 10: Möbius strip

1 (a) The Möbius strip is an object that has always fascinated me.
   (b) Basically, it is a real 3D object with a surface that has only one side.

2 (a) Think about a strip of paper that is twisted and turned, so that the 'top end of the bottom surface' is laid onto the 'top end of the top surface'.
   (b) A line drawn from the seam down the middle will meet back at the seam, but on the other side (i.e. it is one continuous surface).

3 (a) Möbius strips are used for several applications in industry.
   (b) Belt drives and conveyor systems are the most obvious use, where the twist concept reduces wear when compared to a normal belt system.

4 Over the years, I have tried (unsuccessfully) to model a Möbius strip using different CAD software packages, and it is only recently that I have managed to create one, or one that looks reasonable, but you can decide this for yourself.

5 The Möbius strip project requires you to:
   (a) use the given X-section drawing information to create three different Möbius strip models (note that you only really need to create one, then modify/edit as required); and
   (b) create a drawing layout for any one of your models, the views being at your discretion.

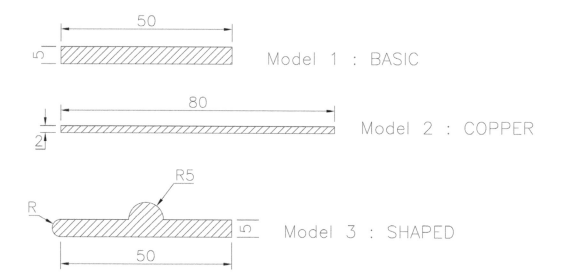

U Final thoughts

1   Hopefully, if you are reading this, you have completed the exercises that appealed to you, and perhaps you even completed all of them.

2   As I stated at the start, my intention was not to produce a 'teaching manual', but a lot of exercises for the user to practise their Inventor skills and abilities.

3   (a) Having completed every exercise in the book, I found that my skill and speed was improving.
      (b) My main problem was completing the various 2D drawings needed to convey the required information for the creation of the various models.
      (c) This I found very tedious and boring.

4   Inventor® is an excellent package and able to compete with similar industrial software (e.g. Solid Works®, CATIA®, Pro/ENGINEER, etc.).

5   If you want to try other model creation, then get any engineering book with sketches and try to create the models displayed.

6   The most difficult models for me to create were:
      (a) P24: the Wall Bracket and the 2D drawing was more difficult than the model.
      (b) S9: Part 1 (Body) of the Small Lathe Tailstock.

7   *Inventor vs AutoCAD*
      (a) All the models in this book were created using Inventor.
      (b) I decided to investigate using AutoCAD and completed several of the more complex models and assemblies.
      (c) Thus, AutoCAD can be used to create the various models in the book.
      (d) The only drawback with AutoCAD is that nuts, bolts, screw threads, etc. are not available and have to be created by the user.

8   (a) The images below display four models created with both Inventor and AutoCAD.
      (b) There is no real difference between the two model 'types' other than threads, etc.
      (c) The four models displayed are:
         •   Location Fixture: P7.
         •   Valve Casing: P11.
         •   Cover: P22.
         •   Back Panel: H3 assembly.

AutoCAD model   *Inventor® model*

AutoCAD model

Inventor® model

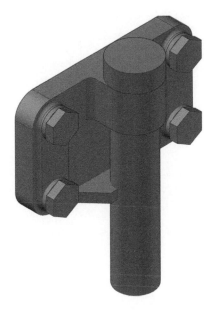

Index

Model index